MW00425179

An Introduction to
ORTHOGONAL
POLYNOMIALS

THEODORE S. CHIHARA
Purdue University, Calumet Campus

DOVER PUBLICATIONS, INC.
Mineola, New York

Copyright

Copyright © 1978 by Theodore S. Chihara
All rights reserved.

Bibliographical Note

This Dover edition, first published in 2011, is an unabridged republication of the work originally published in 1978 by Gordon and Breach, Science Publisher, Inc., New York.

Library of Congress Cataloging-in-Publication Data

Chihara, Theodore S. (Theodore Seio), 1929–
    An introduction to orthogonal polynomials / Theodore S. Chihara. — Dover ed.
        p. cm.
    Originally published: New York : Gordon and Breach, c1978.
    Includes index.
    Summary: "This concise introduction covers general elementary theory related to orthogonal polynomials and assumes only a first undergraduate course in real analysis. Topics include the representation theorem and distribution functions, continued fractions and chain sequences, the recurrence formula and properties of orthogonal polynomials. 1978 edition"—Provided by publisher.
    ISBN-13: 978-0-486-47929-3 (pbk.)
    ISBN-10: 0-486-47929-3 (pbk.)
    1. Functions, Orthogonal. I. Title.

QA404.5C44 2011
515'.55—dc22

2010043412

Manufactured in the United States by Courier Corporation
47929301
www.doverpublications.com

To

A, G, J, and $L_i (i = 1, 2, 3)$

Preface

The subject of orthogonal polynomials is a classical one whose origins can be traced to Legendre's work on planetary motion. With important applications to physics and to probability and statistics and other branches of mathematics, the subject flourished through the first third of this century. After the publication of Szegö's well known treatise on the subject, interest in orthogonal polynomials began to wane as mathematicians turned their attention to increasingly greater abstraction.

Perhaps as a secondary effect of the computer revolution and the heightened activity in approximation theory and numerical analysis, interest in orthogonal polynomials has revived in recent years. Despite this increased attention, courses on the subject remain virtually nonexistent (a situation I feel is both natural and proper). The subject is usually introduced into the curriculum in the midst of a course in approximation theory or methods of mathematical physics. These introductions are always brief and deal almost exclusively with the properties of the classical orthogonal polynomials. On the other hand, more comprehensive books on the subject usually assume a great deal of classical mathematics as background and get rather quickly into material requiring a knowledge of advanced mathematics.

Thus I felt it might be useful to write a book which dealt with the basic general theory and included the necessary background material which is not usually found in the standard mathematics curriculum. Further, in view of the absence of formal courses on the subject, I have tried to write it in a manner suitable for independent study by an advanced undergraduate or beginning graduate student.

The text proper comprises the first four chapters and deals with the general theory of orthogonal polynomials including development of necessary material from the problem of moments and the theory of continued fractions. The only prerequisite for reading this portion of the book is a little elementary linear algebra and a first course in undergraduate real analysis as provided, for example, by the first seven chapters of the text by Rudin [1]. With one exception, everything in the first four chapters is proved assuming no further background. The one exception occurs in §4 of

Chapter IV where a little basic analytic function theory and the Stieltjes-Vitali Theorem are used. No Lebesgue theory is required and even the Riemann-Stieltjes integral is not needed until Chapter II.

Our attention throughout the book is focused on orthogonal polynomials as polynomials. To those who detect a personal bias in favor of recurrence relations, we plead guilty but rationalize that this is one aspect of the theory where a great deal can be developed using only very elementary tools. Thus we do not discuss such important topics as asymptotic behavior, expansion theory, interpolation and approximation theory nor orthogonal polynomials in a complex variable. But here we can take refuge in the fact that existing books contain better treatments of these topics than we could provide anyway.

The exercises form an integral part of the text. Any serious study of orthogonal polynomials must include a familiarity with special functions and especially the so-called classical orthogonal polynomials. While a few examples of specific orthogonal polynomials have been introduced into the text, it seemed undesirable to interrupt the systematic development of the theory with detailed discussions of the properties of specific polynomials. For this reason, examples have been introduced through the exercises in a manner which, it is hoped, will suggest to the reader some of the flavor and methods of the "special functions theory" of orthogonal polynomials.

Chapters V and VI are of a markedly different character from the preceding ones. Here we drop all prerequisite limitations and present summaries of the properties of many of the specific examples that have appeared in the literature. No claim to completeness is made but we believe that a majority of those orthogonal polynomials that have received any kind of attention in the literature have been included. Since detailed, comprehensive treatments of the classical orthogonal polynomials are readily available we have concentrated our discussion of these polynomials on their many unifying properties.

Much of the material of these last two chapters has been scattered through the literature. Bringing them together here should provide the beginner with a view of the state of the special functions aspect of orthogonal polynomials and may also be useful to active workers in the field. Perhaps it will also help reduce the number of times some of these examples will be rediscovered. In this connection, the table of recurrence formulas we have included in an appendix may also be of value.

To complement Chapters I–IV, we have added a Notes section at the end of the book. This contains references, historical remarks and indications of other aspects of the subject and of related material that we could not include because of the limitations imposed by our prerequisites. This should

provide suggestions for further study and an introduction to some areas of
current research. (Some exercises in the text have been included for this
same purpose.) For the last two chapters original sources have been referred
to within the text.

The bibliography consists of items actually referred to in the text and
Notes. While quite extensive, it is certainly far from comprehensive since
we have been content in many cases to refer to standard existing books for
references to original sources. Thus, for example, even such twentieth
century names as Erdös, Pólya and Turán are missing.

The numbering system used in this book is the common one whereby
(3.12) refers to the 12th numbered display in section 3 of the same chapter.
The same scheme is followed for exercises and theorems but not for lemmas
and corollaries. For these latter, we simply make such references as "the
corollary to Theorem 4.1." References to items from a different chapter are
indicated by prefixing the appropriate chapter numeral. Thus II-Theorem
3.5 refers to Theorem 3.5 of Chapter II. We have adopted the practice
introduced by Halmos of indicating the end of a proof by the symbol ∎.

I am pleased to acknowledge various aids and courtesies extended to me
during the writing of this book by Seattle University, the University of
Victoria and, especially, the University of Alberta. I am grateful to
Professor André Yandl for his initial encouragement and to Professor Mary
Turner whose critical reading of Chapter I helped the initial writing phases
considerably. Professor Waleed Al-Salam, who used portions of the manuscript in a seminar and pointed out many errors, was a source of much help
and encouragement. Professor Richard Askey read almost the entire
manuscript with a gimlet eye and made several valuable suggestions which
have been incorporated. In addition, he pointed out numerous misprints
and errors, a couple of which would have proven highly embarrassing had
they appeared in print.

While gratefully acknowledging the help of these colleagues, I also add
the customary disclaimer of any responsibility on their part for any errors
or inadequacies that remain. For these, lacking a co-author to blame, I
reluctantly accept full responsibility.

Contents

CHAPTER I

Elementary Theory of Orthogonal Polynomials

1 Introduction

It is an elementary exercise in calculus to use the trigonometric identity

$$2 \cos m\theta \cos n\theta = \cos(m + n)\theta + \cos(m - n)\theta \qquad (1.1)$$

to obtain the integration formula

$$\int_0^\pi \cos m\theta \cos n\theta\, d\theta = 0, \qquad m \neq n; \qquad m, n = 0, 1, 2, \ldots. \quad (1.2)$$

The fact that the integral in (1.2) vanishes is expressed by saying that $\cos m\theta$ and $\cos n\theta$ are *orthogonal* over the interval $(0, \pi)$ for $m \neq n$. We also say that $\{1, \cos\theta, \cos 2\theta, \ldots, \cos n\theta, \ldots\}$ is an *orthogonal sequence* over $(0, \pi)$.

We observe that the change of variable, $x = \cos\theta$, converts (1.2) into

$$\int_{-1}^1 T_m(x) T_n(x)(1 - x^2)^{-1/2}\, dx = 0, \qquad m \neq n, \quad (1.3)$$

where we have written

$$T_n(x) = \cos n\theta = \cos(n \cos^{-1} x), \qquad -1 \leq x \leq 1. \quad (1.4)$$

We have

$$T_0(x) = 1, \qquad T_1(x) = \cos\theta = x,$$

and by elementary trigonometric identities,

$$T_2(x) = 2x^2 - 1, \qquad T_3(x) = 4x^3 - 3x, \qquad \text{etc.}$$

1

Using the identity (1.1) with $m = 1$, it is an easy proof by induction to show that $T_n(x)$ is a polynomial in x of degree n. These polynomials are called the *Tchebichef polynomials of the first kind*. Because of (1.3) we say that the $T_n(x)$ are *orthogonal polynomials* with respect to $(1 - x^2)^{-1/2}$, $-1 < x < 1$. More precisely, $\{T_n(x)\}_{n=0}^{\infty}$ is an *orthogonal polynomial sequence* with respect to the *weight function* $(1 - x^2)^{-1/2}$ on the interval $(-1, 1)$.

Somewhat more generally, consider a function w which is non-negative and integrable on an interval (a, b). We also assume that $w(x) > 0$ on a sufficiently large subset of (a, b) so that

$$\int_a^b w(x)\, dx > 0$$

(that is, $w(x) > 0$ on a subset of positive Lebesgue measure). In the event that (a, b) is unbounded, we will also have to impose the additional requirement that the "moments"

$$\mu_n = \int_a^b x^n w(x)\, dx, \qquad n = 0, 1, 2, \ldots \tag{1.5}$$

are all finite.

Now if there is a sequence of polynomials $\{P_n(x)\}_{n=0}^{\infty}$, $P_n(x)$ of degree n, such that

$$\int_a^b P_m(x) P_n(x) w(x)\, dx = 0 \qquad \text{for } m \neq n, \tag{1.6}$$

then $\{P_n(x)\}$ is called an *orthogonal polynomial sequence* with respect to the *weight function* w on (a, b).

If we write, for any integrable function f,

$$\mathcal{L}[f] = \int_a^b f(x) w(x)\, dx \tag{1.7}$$

then (1.5) and (1.6) can be written

$$\mathcal{L}[x^n] = \mu_n, \qquad n = 0, 1, 2, \ldots, \tag{1.8}$$

$$\mathcal{L}[P_m(x) P_n(x)] = 0, \qquad m \neq n; \qquad m, n = 0, 1, 2, \ldots. \tag{1.9}$$

We note that $\mathcal{L}$ is linear; that is,

$$\mathcal{L}[af(x) + bg(x)] = a\mathcal{L}[f(x)] + b\mathcal{L}[g(x)], \tag{1.10}$$

for arbitrary constants a and b and integrable functions f and g. Without reference to (1.7), we also observe that (1.8) and (1.10) alone are sufficient to define $\mathcal{L}[\pi(x)]$ for any polynomial $\pi(x)$. Indeed,

$$\mathcal{L}\left\{\sum_{k=0}^{n} c_k x^k\right\} = \sum_{k=0}^{n} c_k \mu_k.$$

The latter observation thus suggests a further generalization. In place of (1.7), consider an arbitrary sequence of real or complex numbers $\{\mu_n\}_{n=0}^{\infty}$. We then use (1.8) and (1.10) to *define* a linear functional $\mathcal{L}$ on the vector space of all polynomials in one real variable.

If there is a sequence $\{P_n(x)\}$ of polynomials satisfying (1.9) and the additional condition

$$\mathcal{L}[P_n^2(x)] \neq 0$$

(which is automatically satisfied if $\mathcal{L}$ is defined by (1.7)), we call it an orthogonal polynomial sequence with respect to $\mathcal{L}$. Not every "moment sequence" $\{\mu_n\}$ will give rise to an orthogonal polynomial sequence so we will be concerned with existence questions.

Before turning to a more formal and detailed study of this generalization, we consider a second example which will illustrate the greater generality we wish to encompass.

Consider the function of two variables x and w, with parameter $a \neq 0$,

$$\begin{aligned} G(x, w) &= e^{-aw}(1 + w)^x \\ &= \sum_{m=0}^{\infty} \frac{(-a)^m w^m}{m!} \sum_{n=0}^{\infty} \binom{x}{n} w^n. \end{aligned} \tag{1.11}$$

Forming the Cauchy product of the two series above, we obtain

$$G(x, w) = \sum_{n=0}^{\infty} P_n(x) w^n \tag{1.12}$$

where

$$P_n(x) = \sum_{k=0}^{n} \binom{x}{k} \frac{(-a)^{n-k}}{(n - k)!}. \tag{1.13}$$

Since

$$\binom{x}{k} = \frac{1}{k!} x(x - 1) \cdots (x - k + 1), \qquad k = 1, 2, \ldots, n,$$

we see that $P_n(x)$ is a polynomial of degree n. $P_n(x)$, or a certain constant multiple of $P_n(x)$, is a *Charlier polynomial* and $G(x, w)$ is called a *generating function* for $\{P_n(x)\}$.

We will now show that these polynomials satisfy an "orthogonality relation" which, at first glance, appears to be of a different type than (1.6). We will proceed formally, leaving it to the reader to supply the justification for the various limit interchanges that are invoked.

Referring to (1.11), we see that

$$a^x G(x, v)G(x, w) = e^{-a(v+w)}[a(1 + v)(1 + w)]^x,$$

hence

$$\sum_{k=0}^{\infty} \frac{a^k G(k, v)G(k, w)}{k!} = e^{-a(v+w)}e^{a(1+v)(1+w)}$$

$$= e^a e^{avw}$$

$$= \sum_{n=0}^{\infty} \frac{e^a a^n (vw)^n}{n!}.$$

At the same time, because of (1.12) we also have

$$\sum_{k=0}^{\infty} \frac{a^k G(k, v)G(k, w)}{k!} = \sum_{k=0}^{\infty} \frac{a^k}{k!} \sum_{m,n=0}^{\infty} P_m(k) P_n(k) v^m w^n$$

$$= \sum_{m,n=0}^{\infty} \sum_{k=0}^{\infty} P_m(k) P_n(k) \frac{a^k}{k!} v^m w^n.$$

Comparing coefficients of $v^m w^n$ in the two resulting series, we conclude

$$\sum_{k=0}^{\infty} P_m(k) P_n(k) \frac{a^k}{k!} = \begin{cases} 0 & \text{if } m \neq n \\ \dfrac{e^a a^n}{n!} & \text{if } m = n. \end{cases} \tag{1.14}$$

We say that $\{P_n(x)\}$ is an orthogonal polynomial sequence with respect to the discrete mass distribution which has mass $a^k/k!$ at the point k ($k = 0, 1, 2, \dots$). If we write

$$\mathcal{L}[x^n] = \sum_{k=0}^{\infty} k^n \frac{a^k}{k!}, \qquad n = 0, 1, 2, \dots, \tag{1.15}$$

and define $\mathcal{L}$ for all polynomials by linearity, then (1.14) can be written

$$\mathcal{L}[P_m(x) P_n(x)] = \frac{e^a a^n}{n!} \delta_{mn}, \qquad m, n = 0, 1, 2, \dots,$$

where δ_{mn} is the "Kronecker delta" defined by

$$\delta_{mn} = \begin{cases} 0 & \text{if } m \neq n \\ 1 & \text{if } m = n. \end{cases} \tag{1.16}$$

Thus both (1.6) and (1.14) can be described in terms of a linear functional $\mathcal{L}$.

Finally we note that if ψ denotes a step function which is constant on each of the open intervals, $(-\infty, 0)$ and $(k, k+1)$ $(k = 0, 1, 2, \ldots)$, and has a jump of magnitude $a^k/k!$ at k $(k = 0, 1, 2, \ldots)$, then (1.14) can be written in terms of a (Riemann) Stieltjes integral as

$$\int_{-\infty}^{\infty} P_m(x) P_n(x) \, d\psi(x) = \frac{e^a a^n}{n!} \delta_{mn}. \tag{1.17}$$

Exercises

1.1 Show that

$$\int_{-1}^{1} T_m(x) T_n(x) (1 - x^2)^{-1/2} \, dx = \begin{cases} \dfrac{\pi}{2} \delta_{mn} & \text{if } n > 0 \\ \pi & \text{if } m = n = 0. \end{cases}$$

1.2 The *Tchebichef polynomials of the second kind* are defined by

$$U_n(x) = \frac{\sin(n+1)\theta}{\sin \theta}, \qquad x = \cos \theta, \qquad n = 0, 1, 2, \ldots .$$

(a) Show that $U_n(x)$ is a polynomial in x of degree n.
(b) Prove that

$$\int_{-1}^{1} U_m(x) U_n(x) (1 - x^2)^{1/2} \, dx = \frac{\pi}{2} \delta_{mn}.$$

1.3 Use DeMoivre's theorem to express $\cos n\theta$ and $\sin(n+1)\theta / \sin \theta$ as polynomials in $\cos \theta$ and thus obtain "explicit" formulas for $T_n(x)$ and $U_n(x)$.

1.4 Let $F(x, w) = e^{-(x-w)^2}$.
(a) Show that

$$\frac{\partial^n F}{\partial w^n}\Big|_{w=0} = (-1)^n D^n e^{-x^2}, \qquad D = \frac{d}{dx},$$

and that the latter is of the form $e^{-x^2} H_n(x)$, where $H_n(x)$ is a polynomial of degree n.
(b) Show that

$$G(x, w) = e^{2xw - w^2} = \sum_{n=0}^{\infty} H_n(x) \frac{w^n}{n!}$$

where $H_n(x) = (-1)^n e^{x^2} D^n e^{-x^2}$ is called the *Hermite polynomial* of degree n.

(c) Using the well known result, $\int_{-\infty}^{\infty} e^{-x^2} dx = \sqrt{\pi}$, prove that

$$\int_{-\infty}^{\infty} G(x, v)G(x, w)e^{-x^2} dx = \sqrt{\pi} e^{2vw}.$$

(d) Finally, prove that

$$\int_{-\infty}^{\infty} H_m(x)H_n(x)e^{-x^2} dx = \sqrt{\pi} \, 2^n n! \, \delta_{mn}.$$

1.5

(a) Show that

$$D^{n+1} e^{-x^2} = -2xD^n e^{-x^2} - 2nD^{n-1} e^{-x^2}$$

and thus obtain the relation

$$H_{n+1}(x) = 2xH_n(x) - 2nH_{n-1}(x).$$

(b) Show that $H'_n(x) = 2xH_n(x) - H_{n+1}(x)$ and conclude that

$$H'_n(x) = 2nH_{n-1}(x).$$

(c) Show that $y = H_n(x)$ satisfies the differential equation

$$y'' - 2xy' + 2ny = 0.$$

1.6 Let $H(x, w) = (1 - 2xw + w^2)^{-1/2} = \sum_{n=0}^{\infty} P_n(x)w^n$.

(a) Show that

$$(1 - 2xw + w^2)\frac{\partial H}{\partial w} - (x - w)H = 0,$$

and deduce that

$$(n + 1)P_{n+1}(x) = (2n + 1)xP_n(x) - nP_{n-1}(x), \qquad n \geq 1.$$

Since $P_0(x) = 1$ and $P_1(x) = x$, it follows that $P_n(x)$ is a polynomial of degree n called the *Legendre polynomial*.

(b) Use the generating function, H, to prove that

$$\int_{-1}^{1} P_m(x)P_n(x) dx = \frac{2}{2n + 1} \delta_{mn}.$$

2 The Moment Functional and Orthogonality

We turn now to a formal discussion of the linear functional $\mathcal{L}$ and the corresponding orthogonal polynomials. Throughout this book, "polynomial" will mean a polynomial with complex coefficients in one variable while "real polynomial" will refer to a polynomial with real coefficients.

DEFINITION 2.1 Let $\{\mu_n\}_{n=0}^{\infty}$ be a sequence of complex numbers and let $\mathcal{L}$ be a complex valued function defined on the vector space of all polynomials by

$$\mathcal{L}[x^n] = \mu_n, \qquad n = 0, 1, 2, \ldots$$

$$\mathcal{L}[\alpha_1 \pi_1(x) + \alpha_2 \pi_2(x)] = \alpha_1 \mathcal{L}[\pi_1(x)] + \alpha_2 \mathcal{L}[\pi_2(x)]$$

for all complex numbers α_i and all polynomials $\pi_i(x)$ $(i = 1, 2)$. Then $\mathcal{L}$ is called the *moment functional* determined by the formal *moment sequence* $\{\mu_n\}$. The number μ_n is called the *moment of order n*.

It follows immediately that if $\pi(x) = \sum_{k=0}^{n} c_k x^k$, then

$$\mathcal{L}[\pi(x)] = \sum_{k=0}^{n} c_k \mu_k.$$

x is always considered a *real* variable in these formulas so we also have

$$\mathcal{L}[\overline{\pi(x)}] = \sum_{k=0}^{n} \overline{c_k} \mu_k,$$

where $\bar{z}$ denotes the complex conjugate of the complex number z.

DEFINITION 2.2   A sequence $\{P_n(x)\}_{n=0}^{\infty}$ is called an *orthogonal polynomial sequence* with respect to a moment functional $\mathcal{L}$ provided for all nonnegative integers m and n,
  (i) $P_n(x)$ is a polynomial of degree n,
  (ii) $\mathcal{L}[P_m(x)P_n(x)] = 0$ for $m \neq n$,
  (iii) $\mathcal{L}[P_n^2(x)] \neq 0$.

"Orthogonal polynomial sequence" will be abbreviated "OPS" and we will use such phrases as "$\{P_n(x)\}$ is an OPS for $\mathcal{L}$." When there is no danger of ambiguity, we will speak loosely of the $P_n(x)$ as "orthogonal polynomials."

If $\{P_n(x)\}$ is an OPS for $\mathcal{L}$ and in addition we also have $\mathcal{L}[P_n^2(x)] = 1$ $(n \geq 0)$, then it will be called an ortho*normal* polynomial sequence. That is, $\{P_n(x)\}$ is an orthonormal polynomial sequence if $P_n(x)$ is a polynomial of degree n and

$$\mathcal{L}[P_m(x)P_n(x)] = \delta_{mn}, \qquad m, n = 0, 1, 2, \ldots \tag{2.1}$$

In the general case, conditions (ii) and (iii) of Definition 2.2 can be replaced by

$$\mathcal{L}[P_m(x)P_n(x)] = K_n \delta_{mn}, \qquad K_n \neq 0. \tag{2.2}$$

(Here, δ_{mn} is Kronecker's delta defined by (1.16).)

It follows from (i) and (iii) of the definition that if an OPS for $\mathcal{L}$ exists, then

$$\mu_0 \neq 0, \qquad P_0(x) \neq 0.$$

Thus no OPS can exist if $\mathcal{L}[1] = 0$. Somewhat less trivially it is easy to show that no OPS can exist if, for example, $\mu_0 = \mu_1 = \mu_2 = 1$. For in this case, we would have

$$P_0(x) = a \neq 0, \qquad P_1(x) = bx + c, \qquad b \neq 0,$$

and (ii) requires

$$\mathcal{L}[P_0(x)P_1(x)] = a(b\mu_1 + c\mu_0) = a(b + c) = 0.$$

Thus we must have $b = -c$ and this yields

$$\mathcal{L}[P_1^2(x)] = b^2(\mu_2 - 2\mu_1 + \mu_0) = 0.$$

Before taking up existence questions in earnest, however, we first note some equivalents to Definition 2.2.

THEOREM 2.1 Let $\mathcal{L}$ be a moment functional and let $\{P_n(x)\}$ be a sequence of polynomials. Then the following are equivalent:
  (a) $\{P_n(x)\}$ is an OPS with respect to $\mathcal{L}$;
  (b) $\mathcal{L}[\pi(x)P_n(x)] = 0$ for every polynomial $\pi(x)$ of degree $m < n$ while $\mathcal{L}[\pi(x)P_n(x)] \neq 0$ if $m = n$;
  (c) $\mathcal{L}[x^m P_n(x)] = K_n \delta_{mn}$ where $K_n \neq 0$, $m = 0, 1, \ldots, n$.

Proof Let $\{P_n(x)\}$ be an OPS for $\mathcal{L}$. Since each $P_k(x)$ is of degree k, it is clear that $\{P_0(x), P_1(x), \ldots, P_m(x)\}$ is a basis for the vector subspace of polynomials of degree at most m. Thus if $\pi(x)$ is a polynomial of degree m, there exist constants c_k such that

$$\pi(x) = \sum_{k=0}^{m} c_k P_k(x), \qquad c_m \neq 0.$$

By the linearity of $\mathcal{L}$,

$$\mathcal{L}[\pi(x)P_n(x)] = \sum_{k=0}^{m} c_k \mathcal{L}[P_k(x)P_n(x)] = 0 \qquad \text{if } m < n$$

$$= c_n \mathcal{L}[P_n^2(x)] \qquad \text{if } m = n.$$

Thus (a) $\Rightarrow$ (b). Since trivially (b) $\Rightarrow$ (c) $\Rightarrow$ (a), this completes the proof. ∎

THEOREM **2.2** Let $\{P_n(x)\}$ be an OPS with respect to $\mathcal{L}$. Then for every polynomial $\pi(x)$ of degree n,

$$\pi(x) = \sum_{k=0}^{n} c_k P_k(x)$$

where

$$c_k = \frac{\mathcal{L}[\pi(x)P_k(x)]}{\mathcal{L}[P_k^2(x)]}, \qquad k = 0, 1, \ldots, n. \tag{2.3}$$

Proof As noted previously, if $\pi(x)$ is a polynomial of degree n, then there are constants c_k such that

$$\pi(x) = \sum_{k=0}^{n} c_k P_k(x).$$

Multiplying both sides of this equation by $P_m(x)$ and applying $\mathcal{L}$ we obtain

$$\mathcal{L}[\pi(x)P_m(x)] = \sum_{k=0}^{n} c_k \mathcal{L}[P_k(x)P_m(x)] = c_m \mathcal{L}[P_m^2(x)].$$

Since $\mathcal{L}[P_m^2(x)] \neq 0$, (2..3) follows. ∎

COROLLARY If $\{P_n(x)\}$ is an OPS for $\mathcal{L}$, then each $P_n(x)$ is uniquely determined up to an arbitrary non-zero factor. That is, if $\{Q_n(x)\}$ is also an OPS for $\mathcal{L}$, then there are constants $c_n \neq 0$ such that

$$Q_n(x) = c_n P_n(x), \qquad n = 0, 1, 2, \ldots. \tag{2.4}$$

Proof If $\{Q_n(x)\}$ is an OPS for $\mathcal{L}$, then by Theorem 1.1,

$$\mathcal{L}[P_k(x)Q_n(x)] = 0 \qquad \text{for } k < n.$$

Thus taking $\pi(x) = Q_n(x)$ in Theorem 2.2, we get (2.4). ∎

It is clear that if $\{P_n(x)\}$ is an OPS for $\mathcal{L}$, then so is $\{c_n P_n(x)\}$ for every sequence of non-zero constants c_n. The above Corollary shows conversely that an OPS $\{P_n(x)\}$ is uniquely determined if it satisfies an additional condition that fixes the leading coefficient (the coefficient of x^n) of each $P_n(x)$.

The simplest and most direct method of singling out a particular OPS for a given moment functional is to specify explicitly the value of each leading coefficient. We will usually "standardize" by requiring that each $P_n(x)$ be a monic polynomial—that is, that $P_n(x)$ have 1 as its leading coefficient. An OPS in which each $P_n(x)$ is monic will be referred to as a *monic OPS*.

Clearly if $\{P_n(x)\}$ is an OPS and k_n denotes the leading coefficient of $P_n(x)$, then

$$\hat{P}_n(x) = k_n^{-1} P_n(x)$$

yields the corresponding monic OPS, $\{\hat{P}_n(x)\}$.

On the other hand,

$$p_n(x) = \{\mathcal{L}[P_n^2(x)]\}^{-1/2} P_n(x)$$

yields $\{p_n(x)\}$ as a corresponding orthonormal polynomial sequence. In general, the square roots appearing here will not be real. However, in the most important occurrences of orthogonal polynomials we will have $\mathcal{L}[P_n^2(x)] > 0$ and in this case $p_n(x)$ can be uniquely determined by the usual additional requirement that its leading coefficient be positive.

Finally, we note the obvious fact that if $\{P_n(x)\}$ is an OPS for $\mathcal{L}$, then $\{P_n(x)\}$ is also an OPS for every moment functional $\mathcal{L}'$ such that for some fixed constant, $c \neq 0$,

$$\mathcal{L}'[x^n] = c\mathcal{L}[x^n], \qquad n = 0, 1, 2, \ldots.$$

Exercises

2.1 Let $\mathcal{L}[x^n] = a^n$ ($n \geq 0$). Show that no OPS for $\mathcal{L}$ exists.

2.2 Let $P_n(x) = x^n$ ($n \geq 0$). Show that $\{P_n(x)\}$ is not an OPS.

2.3 Let $\mathcal{L}$ be a moment functional such that an OPS for $\mathcal{L}$ exists. Let C_n be arbitrary non-zero numbers and show that each of the following uniquely determines a corresponding OPS, $\{P_n(x)\}$.
   (a) $P_n(x_0) = C_n$ where x_0 is not a zero of any $P_k(x)$;
   (b) $\mathcal{L}[x^n P_n(x)] = C_n$;
   (c)

$$\lim_{x \to 0} x^n P_n(1/x) = C_n, \qquad n = 0, 1, 2, \ldots.$$

2.4 Given a moment functional $\mathcal{L}$ with moment sequence $\{\mu_n\}$, let $\mathfrak{M}$ be defined by

$$\mathfrak{M}[x^n] = \mathcal{L}[(ax + b)^n], \qquad a \neq 0, \qquad n = 0, 1, 2, \ldots.$$

If $\{P_n(x)\}$ is the monic OPS for $\mathcal{L}$, find the monic OPS for $\mathfrak{M}$.

2.5 Find explicitly $\{P_n(x)\}$ such that
  (a) $\int_0^1 P_m(x) P_n(x) x^{-1/2} (1-x)^{-1/2} dx = K_n \delta_{mn}$ where $K_n = \pi/2$ for $n > 0$ and $K_0 = \pi$;
  (b) $\int_{-\infty}^{\infty} P_m(x) P_n(x) e^{-x^2/2} dx = \sqrt{2\pi}\, n!\, \delta_{mn}$;
  (c)

$$\int_0^1 P_m(x) P_n(x)\, dx = \frac{1}{2n+1} \delta_{mn}.$$

2.6 If $\{P_n(x)\}$ is an OPS for $\mathcal{L}$ and if k_n denotes the leading coefficient of $P_n(x)$, prove that $\mathcal{L}[P_m(x)\overline{P_n(x)}] = L_n \delta_{mn}$, where $L_n = \overline{k_n}\mathcal{L}[P_n^2(x)]/k_n$ ($n \geq 0$).

2.7 With the notation of Theorem 2.2, prove that for every polynomial $\pi(x)$ of degree n,

$$\mathcal{L}\{|\pi(x)|^2\} = \sum_{k=0}^{n} |c_k|^2 \mathcal{L}\{|P_k(x)|^2\}.$$

3 Existence of OPS

In order to discuss existence theorems for OPS, we introduce the determinants

$$\Delta_n = \det(\mu_{i+j})_{i,j=0}^{n} = \begin{vmatrix} \mu_0 & \mu_1 & \cdots & \mu_n \\ \mu_1 & \mu_2 & \cdots & \mu_{n+1} \\ \cdot & \cdot & \cdots & \cdot \\ \cdot & \cdot & \cdots & \cdot \\ \mu_n & \mu_{n+1} & \cdots & \mu_{2n} \end{vmatrix} \qquad (3.1)$$

THEOREM 3.1   Let $\mathcal{L}$ be a moment functional with moment sequence $\{\mu_n\}$. A necessary and sufficient condition for the existence of an OPS for $\mathcal{L}$ is

$$\Delta_n \neq 0, \qquad n = 0, 1, 2, \ldots .$$

Proof Write

$$P_n(x) = \sum_{k=0}^{n} c_{nk} x^k.$$

Recalling Theorem 2.1, we observe that the orthogonality conditions

$$\mathcal{L}[x^m P_n(x)] = \sum_{k=0}^{n} c_{nk}\mu_{k+m} = K_n \delta_{mn}, \qquad K_n \neq 0, \qquad m \leq n, \quad (3.2)$$

are equivalent to the system

$$
\begin{bmatrix}
\mu_0 & \mu_1 & \cdots & \mu_n \\
\mu_1 & \mu_2 & \cdots & \mu_{n+1} \\
\cdot & \cdot & \cdots & \cdot \\
\cdot & \cdot & \cdots & \cdot \\
\mu_{2n} & \cdots & & \mu_n
\end{bmatrix}
\begin{bmatrix}
c_{no} \\
c_{n1} \\
\cdot \\
\cdot \\
c_{nn}
\end{bmatrix}
=
\begin{bmatrix}
0 \\
0 \\
\cdot \\
\cdot \\
K_n
\end{bmatrix}
\qquad (3.3)
$$

Now if an OPS for $\mathcal{L}$ exists, it is uniquely determined by the constants K_n in (3.2) (Ex. 2.3). It then follows that (3.2) has a unique solution so that $\Delta_n \neq 0 \ (n \geq 0)$.

Conversely, if $\Delta_n \neq 0$, then for arbitrary $K_n \neq 0$, (3.3) has a unique solution so $P_n(x)$ satisfying (3.2) exists. We also have

$$c_{nn} = \frac{K_n \Delta_{n-1}}{\Delta_n} \neq 0, \qquad n \geq 1, \qquad (3.4)$$

which is valid for $n = 0$ also if we define $\Delta_{-1} = 1$. It follows that $P_n(x)$ is of degree n, hence $\{P_n(x)\}$ is an OPS for $\mathcal{L}$. ∎

Formula (3.4) is sufficiently useful to note it formally.

THEOREM 3.2 Let $\{P_n(x)\}$ be an OPS for $\mathcal{L}$. Then for any polynomial $\pi_n(x)$ of degree n,

$$\mathcal{L}[\pi_n(x) P_n(x)] = a_n \mathcal{L}[x^n P_n(x)] = \frac{a_n k_n \Delta_n}{\Delta_{n-1}}, \qquad \Delta_{-1} = 1, \qquad (3.5)$$

where a_n denotes the leading coefficient of $\pi_n(x)$ and k_n denotes the leading coefficient of $P_n(x)$.

Proof Writing

$$\pi_n(x) = a_n x^n + \pi_{n-1}(x),$$

where $\pi_{n-1}(x)$ is a polynomial of degree $n - 1$, we have

$$\mathcal{L}[\pi_n(x) P_n(x)] = a_n \mathcal{L}[x^n P_n(x)] + \mathcal{L}[\pi_{n-1}(x) P_n(x)]$$
$$= a_n \mathcal{L}[x^n P_n(x)].$$

Thus (3.5) follows from (3.4) with $k_n = c_{nn}$. ∎

In the most important occurrences of orthogonal polynomials, the moment functional $\mathcal{L}$ is defined by a non-negative weight function as in (1.7). Somewhat more generally, as in (1.17), $\mathcal{L}$ is frequently defined in terms of a Stieltjes integral:

$$\mathcal{L}[x^n] = \int_{-\infty}^{\infty} x^n \, d\psi(x), \tag{3.6}$$

where ψ is a bounded, non-decreasing function such that the set

$$\mathfrak{S}(\psi) = \{x \mid \psi(x + \epsilon) - \psi(x - \epsilon) > 0 \qquad \text{for } \epsilon > 0\}$$

is an infinite set.

In such cases, if $\pi(x)$ is a polynomial not identically zero which is non-negative for all real x, then $\mathcal{L}[\pi(x)] > 0$. It will be shown in Chapter II that this property characterizes moment functionals that can be represented as in (3.6). For now, we make the following definition.

DEFINITION 3.1 A moment functional $\mathcal{L}$ is called *positive-definite* if $\mathcal{L}[\pi(x)] > 0$ for every polynomial $\pi(x)$ that is not identically zero and is non-negative for all real x.

If $\mathcal{L}$ is positive-definite, it follows immediately that

$$\mu_{2k} = \mathcal{L}[x^{2k}] > 0,$$

Since

$$0 < \mathcal{L}[(x + 1)^{2n}] = \sum_{k=0}^{2n} \binom{2n}{k} \mu_{2n-k},$$

it follows by induction that μ_{2k+1} is real.

When $\mathcal{L}$ is positive-definite, a step-by-step method of constructing a corresponding ortho*normal* polynomial sequence can be described. Known as the *Gram-Schmidt process* it produces real orthonormal polynomials as follows.

First define

$$p_0(x) = \mu_0^{-1/2}$$

so that $\mathcal{L}[p_0^2(x)] = \mu_0^{-1} \mathcal{L}[1] = 1$.

Next let

$$P_1(x) = x - a p_0(x).$$

Then

$$\mathcal{L}[\,p_0(x)P_1(x)] = \mathcal{L}[xp_0(x)] - a\mathcal{L}[\,p_0^2(x)] = 0$$

provided we choose $a = \mathcal{L}[xp_0(x)]$. With this choice of a, then, we define

$$p_1(x) = \{\mathcal{L}[P_1^2(x)]\}^{-1/2}P_1(x)$$

and observe that

$$\mathcal{L}[\,p_0(x)p_1(x)] = 0, \qquad \mathcal{L}[\,p_1^2(x)] = 1.$$

Note also that $p_0(x)$ and $p_1(x)$ are both real polynomials.

In general, suppose $p_0(x)$, $p_1(x)$, ..., $p_n(x)$ have been constructed such that each $p_i(x)$ is a real polynomial of degree i and

$$\mathcal{L}[\,p_i(x)p_j(x)] = \delta_{ij}, \qquad i, j = 0, 1, \ldots, n.$$

We then define $P_{n+1}(x)$ by

$$P_{n+1}(x) = x^{n+1} - \sum_{k=0}^{n} a_k p_k(x), \qquad a_k = \mathcal{L}[x^{n+1}p_k(x)].$$

Then $P_{n+1}(x)$ is a real polynomial of degree $n + 1$ and we have

$$\mathcal{L}[\,p_j(x)P_{n+1}(x)] = \mathcal{L}[x^{n+1}p_j(x)] - \sum_{k=0}^{n} a_k \mathcal{L}[\,p_j(x)p_k(x)]$$

$$= a_j - a_j = 0.$$

We therefore set

$$p_{n+1}(x) = \{\mathcal{L}[P_{n+1}^2(x)]\}^{-1/2}P_{n+1}(x).$$

Then $p_{n+1}(x)$ is real and

$$\mathcal{L}[\,p_j(x)p_{n+1}(x)] = \delta_{j,n+1}, \qquad j = 0, 1, \ldots, n + 1.$$

By induction, it follows that a real orthonormal polynomial sequence for $\mathcal{L}$ exists.

For convenience of reference, we summarize the above formally.

THEOREM 3.3 Let $\mathcal{L}$ be positive-definite. Then $\mathcal{L}$ has real moments and a corresponding OPS consisting of real polynomials exists.

We next relate the concept of positive-definite moment functionals to the determinants (3.1). We will need the following classical result characterizing non-negative polynomials.

LEMMA   Let $\pi(x)$ be a polynomial that is non-negative for all real x. Then there are real polynomials $p(x)$ and $q(x)$ such that

$$\pi(x) = p^2(x) + q^2(x).$$

Proof   If $\pi(x) \geqq 0$ for real x, then $\pi(x)$ is a real polynomial so its real zeros have even multiplicity and its non-real zeros occur in conjugate pairs. Thus we can write

$$\pi(x) = r^2(x) \prod_{k=1}^{m} (x - \alpha_k - \beta_k i)(x - \alpha_k + \beta_k i),$$

where $r(x)$ is a real polynomial, α_k and β_k are real numbers.
Writing

$$\prod_{k=1}^{m} (x - \alpha_k - \beta_k i) = A(x) + iB(x)$$

where $A(x)$ and $B(x)$ are real polynomials, we get

$$\pi(x) = r^2(x)[A^2(x) + B^2(x)]. \quad \blacksquare$$

THEOREM 3.4 $\mathcal{L}$ is positive-definite if and only if its moments are all real and $\Delta_n > 0$ $(n \geqq 0)$.

Proof   Let μ_n be real, $\Delta_n > 0$ $(n \geqq 0)$. According to Theorem 3.1, an OPS $\{P_n(x)\}$ exists for $\mathcal{L}$. We can assume without loss of generality that $P_n(x)$ is monic. By Theorem 3.2 we have

$$\mathcal{L}[P_n^2(x)] = \Delta_n/\Delta_{n-1} > 0.$$

Referring to (3.3), we see that $P_n(x)$ is real. Thus if $p(x)$ is a real polynomial of degree m, then

$$p(x) = \sum_{k=0}^{m} a_k P_k(x),$$

where the a_k are all real and $a_m \neq 0$. Therefore

$$\mathcal{L}[p^2(x)] = \sum_{j,k=0}^{m} a_j a_k \mathcal{L}[P_j(x) P_k(x)] = \sum_{k=0}^{m} a_k^2 \mathcal{L}[P_k^2(x)] > 0.$$

It thus follows from the lemma that $\mathcal{L}$ is positive-definite.

Conversely, suppose $\mathcal{L}$ is positive-definite so that by Theorem 3.3 its moments are real and an OPS, $\{P_n(x)\}$, for $\mathcal{L}$ exists. Again supposing for the sake of simplicity that $\{P_n(x)\}$ is monic, we have as before

$$0 < \mathcal{L}[P_n^2(x)] = \Delta_n/\Delta_{n-1}, \qquad n \geqq 0.$$

Since $\Delta_{-1} = 1$, it follows that $\Delta_n > 0$ $(n \geqq 0)$. ∎

COROLLARY   Let $\{P_n(x)\}$ be an OPS for $\mathcal{L}$. If $P_n(x)$ is real and $\mathcal{L}[P_n^2(x)] > 0$, then $\mathcal{L}$ is positive-definite.

In the theory of real symmetric matrices, it is shown that the positivity of Δ_n is equivalent to the fact that

$$\sum_{i,j=0}^{n} \mu_{i+j} x_i x_j > 0 \qquad (3.7)$$

for all real x_i such that $(x_0, x_1, \ldots, x_n) \neq (0, 0, \ldots, 0)$ (see Ex. 3.7). Since quadratic forms with this property have long been called positive-definite, the name has been applied to the corresponding moment functional. By analogy and in view of Theorem 3.2, we introduce the following terminology.

DEFINITION 3.2 $\mathcal{L}$ is called *quasi-definite* if and only if $\Delta_n \neq 0$ for $n \geqq 0$.

REMARK   If $\mathcal{L}$ is positive-definite and we define

$$(p, q) = \mathcal{L}[p(x)\overline{q(x)}] \qquad (3.8)$$

for all polynomials $p(x)$ and $q(x)$, then it is easily verified that (3.8) defines an *inner product* on the vector space of all polynomials in one real variable. In particular, if $\{P_n(x)\}$ is an OPS for $\mathcal{L}$, then

$$(P_m, P_n) = \mathcal{L}[P_m(x)\overline{P_n(x)}] = \mathcal{L}[P_m(x)P_n(x)] = 0, \qquad m \neq n.$$

Thus our definition of orthogonality for $\{P_n(x)\}$ is consistent with the usual definition of orthogonality in an inner product space. Hence in the positive-definite case, the well known theory of inner product spaces could be applied to our study of orthogonal polynomials. In particular, Theorem 2.2 gives the usual "Fourier coefficients" and the Gram-Schmidt process is the usual one in inner product spaces.

While our definition of orthogonality in the general quasi-definite case is a natural enough analogue of the positive-definite case, it should be noted that when the moments are non-real, our definition bears no relation to the standard concept of orthogonality of polynomials in a complex variable.

Exercises

Throughout these exercises, $\mathcal{L}$ denotes a quasi-definite moment functional with moment sequence $\{\mu_n\}$. Δ_n denotes the determinant (3.1).

3.1 Let

$$D_n(x) = \begin{vmatrix} \mu_0 & \mu_1 & \cdots & \mu_n \\ \mu_1 & \mu_2 & \cdots & \mu_{n+1} \\ \cdot & \cdot & \cdots & \cdot \\ \cdot & \cdot & \cdots & \cdot \\ \mu_{n-1} & \mu_n & \cdots & \mu_{2n-1} \\ 1 & x & \cdots & x^n \end{vmatrix}$$

(a) Show that $\{(\Delta_{n-1})^{-1} D_n(x)\}$ is the monic OPS for $\mathcal{L}$.
(b) Determine a corresponding orthonormal polynomial sequence.

3.2 If $\mathfrak{M}$ is defined by $\mathfrak{M}[x^n] = \mathcal{L}[(ax + b)^n]$ $(a \neq 0, n \geqq 0)$, then $\mathfrak{M}$ is quasi-definite. It is positive-definite if and only if $\mathcal{L}$ is positive-definite.

3.3 Prove that if $\pi_n(x)$ is a polynomial of degree n, $n = 0, 1, 2, \ldots$, and $\mathcal{L}[\pi_m(x)\pi_n(x)] = 0$ for $m \neq n$, then $\{\pi_n(x)\}$ is an OPS for $\mathcal{L}$.

3.4 Let $\{P_n(x)\}$ be the monic OPS for $\mathcal{L}$. Prove that if $\mathcal{L}$ is positive-definite, the following *extremal property* holds:

$$\mathcal{L}[P_n^2(x)] < \mathcal{L}\{|\pi(x)|^2\}$$

for every monic polynomial, $\pi(x) \neq P_n(x)$, of degree n.

3.5 Suppose $\mathcal{L}$ is defined by (1.7) and let $\{p_n(x)\}$ denote the corresponding sequence of orthonormal polynomials. If f is any complex-valued function such that

$$\mathcal{L}\{|f(x)|^2\} = \int_a^b |f(x)|^2 w(x)\, dx$$

exists, let

$$s_n(x) = s_n(f; x) = \sum_{k=0}^n c_k p_k(x), \qquad c_k = \mathcal{L}[f(x)p_k(x)].$$

Prove that $\mathcal{L}\{|f(x) - s_n(x)|^2\} < \mathcal{L}\{|f(x) - t_n(x)|^2\}$ for every $t_n(x) = \sum_{k=0}^n b_k p_k(x)$ such that $b_i \neq c_i$ for at least one i. [*Least squares property* of the Fourier coefficients].

3.6 With the notation of the preceding exercise, prove *Bessel's inequality*: $\mathcal{L}\{|f(x)|^2\} \geqq \sum_{k=0}^\infty |c_k|^2$.

3.7 Without reference to the theory of matrices, prove that the quadratic forms (3.7) are all positive if and only if $\Delta_n > 0$ $(n \geqq 0)$.

3.8 Let $\mathfrak{L}$ be quasi-definite. Prove that if $\pi(x)$ is a polynomial such that $\mathfrak{L}[\pi(x)x^k] = 0$ for $k = 0, 1, 2, \ldots$, then $\pi(x) \equiv 0$.

3.9 Prove that if the Δ_n are all positive, the moments are all real.

4 The Fundamental Recurrence Formula

One of the most important characteristics of orthogonal polynomials is the fact that any three consecutive polynomials are connected by a very simple relation which we now derive.

THEOREM **4.1** Let $\mathfrak{L}$ be a quasi-definite moment functional and let $\{P_n(x)\}$ be the corresponding monic OPS. Then there exist constants c_n and $\lambda_n \neq 0$ such that

$$P_n(x) = (x - c_n)P_{n-1}(x) - \lambda_n P_{n-2}(x), \qquad n = 1, 2, 3, \ldots, \quad (4.1)$$

where we define $P_{-1}(x) = 0$.

Moreover, if $\mathfrak{L}$ is positive-definite, then c_n is real and $\lambda_{n+1} > 0$ for $n \geqq 1$ (λ_1 is arbitrary).

Proof Since $xP_n(x)$ is a polynomial of degree $n + 1$, we can write

$$xP_n(x) = \sum_{k=0}^{n+1} a_{nk} P_k(x), \qquad a_{nk} = \frac{\mathfrak{L}[xP_n(x)P_k(x)]}{\mathfrak{L}[P_k^2(x)]}.$$

But $xP_k(x)$ is a polynomial of degree $k + 1$ so that $a_{nk} = 0$ for $0 \leqq k < n - 1$. Further, $xP_n(x)$ is monic so $a_{n,n+1} = 1$. Thus

$$xP_n(x) = P_{n+1}(x) + a_{nn}P_n(x) + a_{n,n-1}P_{n-1}(x), \qquad n \geqq 1.$$

Replacing n by $n - 1$, the latter can be written in the form

$$xP_{n-1}(x) = P_n(x) + c_n P_{n-1}(x) + \lambda_n P_{n-2}(x), \qquad n \geqq 2,$$

and this is equivalent to (4.1) for $n \geqq 2$. But (4.1) is valid also for $n = 1$ if we define $P_{-1}(x) = 0$ and choose $c_1 = -P_1(0)$ (λ_1 is arbitrary).

Next from (4.1) we obtain

$$\mathfrak{L}[x^{n-2}P_n(x)] = \mathfrak{L}[x^{n-1}P_{n-1}(x)] - c_n\mathfrak{L}[x^{n-2}P_{n-1}(x)] - \lambda_n\mathfrak{L}[x^{n-2}P_{n-2}(x)],$$

$$0 = \mathfrak{L}[x^{n-1}P_{n-1}(x)] - \lambda_n\mathfrak{L}[x^{n-2}P_{n-2}(x)].$$

By Theorem 3.2, we obtain for $n \geq 1$,

$$\lambda_{n+1} = \frac{\mathcal{L}[x^n P_n(x)]}{\mathcal{L}[x^{n-1} P_{n-1}(x)]} = \frac{\Delta_{n-2} \Delta_n}{\Delta_{n-1}^2} \qquad (\Delta_{-1} = 1).$$

It follows that $\lambda_n \neq 0$ if $\mathcal{L}$ is quasi-definite while if $\mathcal{L}$ is positive-definite, then $\lambda_n > 0$ ($n \geq 2$). Finally, the reality of the c_n follows from the reality of the $P_k(x)$. ∎

THEOREM **4.2** With reference to the recurrence formula (4.1), the following are valid for $n \geq 1$.
(a)

$$\lambda_{n+1} = \frac{\mathcal{L}[P_n^2(x)]}{\mathcal{L}[P_{n-1}^2(x)]} = \frac{\Delta_{n-2} \Delta_n}{\Delta_{n-1}^2}.$$

(b) $\mathcal{L}[P_n^2(x)] = \lambda_1 \lambda_2 \cdots \lambda_{n+1}$ provided we define $\lambda_1 = \mu_0 = \Delta_0$.
(c)

$$c_n = \frac{\mathcal{L}[x P_{n-1}^2(x)]}{\mathcal{L}[P_{n-1}^2(x)]}.$$

(d) The coefficient of x^{n-1} in $P_n(x)$ is $-(c_1 + c_2 + \cdots + c_n)$.

Proof Formula (a) was obtained in the proof of Theorem 4.1 while (b) follows immediately from (a).

To obtain (c), we multiply both sides of (4.1) by $P_{n-1}(x)$ and apply $\mathcal{L}$.

Finally, if d_n denotes the coefficient of x^{n-1} in $P_n(x)$, then comparison of the coefficients of x^{n-1} on both sides of (4.1) shows that $d_n = d_{n-1} - c_n$ from which (d) follows. ∎

In the event $\{P_n(x)\}$ is not monic, then it will satisfy a recurrence formula of the form

$$P_{n+1}(x) = (A_n x + B_n) P_n(x) - C_n P_{n-1}(x), \qquad n \geq 0. \qquad (4.2)$$

(In the non-monic case, it is customary to write the recurrence with the highest index shifted to $n + 1$.) Writing $P_n(x) = k_n \hat{P}_n(x)$ where $\hat{P}_n(x)$ is monic, we find

$$A_n = k_n^{-1} k_{n+1}, \qquad B_n = -c_{n+1} k_n^{-1} k_{n+1}, \qquad C_n = \lambda_{n+1} k_{n-1}^{-1} k_{n+1}, \qquad n \geq 0, \qquad (4.3)$$

where $k_{-1} = 1$, and c_n, λ_n are given by Theorem 4.2 in terms of $\{\hat{P}_n(x)\}$.

In particular, $A_n \neq 0$, $C_n \neq 0$, while the condition for positive-definiteness of $\mathcal{L}$ takes the form

$$C_n A_n A_{n-1} > 0, \qquad n \geq 1. \qquad (4.4)$$

In many cases, the coefficients in the recurrence formulas can be obtained directly from known properties of the specific OPS in question. For example, for the Tchebichef polynomials of the first kind (see §1), we can use the trigonometric identity (1.1) with $m = 1$,

$$\cos(n + 1)\theta + \cos(n - 1)\theta = 2 \cos n\theta \cos \theta, \qquad n \geq 1,$$

to obtain

$$T_{n+1}(x) = 2xT_n(x) - T_{n-1}(x), \qquad n \geq 1. \qquad (4.5)$$

Noting that $T_0(x) = 1$, $T_1(x) = x$, it follows inductively from (4.5) that the leading coefficient of $T_n(x)$ is 2^{n-1}. Thus the corresponding monic polynomials are

$$\hat{T}_0(x) = T_0(x), \qquad \hat{T}_n(x) = 2^{1-n}T_n(x), \qquad n \geq 1,$$

and the recurrence formula (4.1) takes the form

$$\hat{T}_n(x) = x\hat{T}_{n-1}(x) - \frac{1}{4}\hat{T}_{n-2}(x), \qquad n \geq 3,$$

$$\hat{T}_2(x) = x\hat{T}_1(x) - \frac{1}{2}\hat{T}_0(x).$$

We note that the coefficient corresponding to c_n in (4.1) is 0. This fact is obviously related to the fact that $T_n(x)$ is an even or odd function as n is even or odd, which in turn is related to the fact that the corresponding moment functional has all of its moments of odd order equal to 0. We generalize these observations.

DEFINITION 4.1 A moment functional is called *symmetric* if all of its moments of odd order are 0.

If $\mathcal{L}$ is defined in terms of a weight function as in (1.7), then it will be symmetric if $a = -b$ and w is an even function on $(-b, b)$. Such is the case, for example, with the Tchebichef polynomials, the Hermite polynomials and the Legendre polynomials mentioned in §1.

THEOREM 4.3   Let $\{P_n(x)\}$ be the monic OPS with respect to a quasi-definite moment functional $\mathcal{L}$. Then the following are equivalent.

(a) $\mathcal{L}$ is symmetric.

(b) $P_n(-x) = (-1)^n P_n(x)$, $n \geq 0$.

(c) In the corresponding recurrence formula (4.1), $c_n = 0 \, (n \geq 1)$.

Proof   We will prove (a) $\Leftrightarrow$ (b) $\Leftrightarrow$ (c).

If $\mathcal{L}$ is symmetric, then $\mathcal{L}[\pi(-x)] = \mathcal{L}[\pi(x)]$ for any polynomial $\pi(x)$. Thus

$$\mathcal{L}[P_m(-x)P_n(-x)] = \mathcal{L}[P_m(x)P_n(x)].$$

Thus by the Corollary to Theorem 2.2, $P_n(-x) = a_n P_n(x)$ where a_n is a constant. Comparison of leading coefficients shows that $a_n = (-1)^n$.

Conversely, if $P_n(-x) = (-1)^n P_n(x)$, then $P_n(x)$ contains only odd powers of x when n is odd. Thus

$$\mathcal{L}[P_1(x)] = \mu_1 = 0, \qquad \mathcal{L}[P_3(x)] = \mu_3 = 0, \qquad \text{etc.,}$$

and it follows by induction that $\mu_{2n+1} = 0 \, (n \geq 0)$.

Turning to the recurrence (4.1), we observe that $(-1)^n P_n(-x) = Q_n(x)$ satisfies (see also Ex. 4.4)

$$Q_n(x) = (x + c_n)Q_{n-1}(x) - \lambda_n Q_{n-2}(x), \qquad n \geq 1.$$

Therefore, if $Q_n(x) = P_n(x)$, then subtracting the above equation from (4.1) yields $2c_n P_{n-1}(x) = 0$, whence $c_n = 0$ for $n \geq 1$.

Conversely, if $c_n = 0$ in (4.1) for $n \geq 1$, then $\{Q_n(x)\}$ satisfies the same recurrence formula as $\{P_n(x)\}$. Since $Q_{-1}(x) = P_{-1}(x)$ and $Q_0(x) = P_0(x)$, it follows that $Q_n(x) = P_n(x)$ for all n.   ■

We next take up the very important converse to Theorem 4.1. This theorem which states that any polynomial sequence that satisfies a recurrence of the form (4.1) is an OPS was first announced by J. Favard in 1935. It was apparently discovered at about the same time independently by J. Shohat and I. Natanson. The result is actually contained implicitly in earlier known results from the theory of continued fractions and a form of it goes back to Stieltjes. We will however refer to it as Favard's theorem.

THEOREM 4.4   (Favard's Theorem). Let $\{c_n\}_{n=1}^{\infty}$ and $\{\lambda_n\}_{n=1}^{\infty}$ be arbitrary sequences of complex numbers and let $\{P_n(x)\}_{n=0}^{\infty}$ be defined by the recurrence formula

$$P_n(x) = (x - c_n)P_{n-1}(x) - \lambda_n P_{n-2}(x), \qquad n = 1, 2, 3, \ldots,$$
$$P_{-1}(x) = 0, \qquad P_0(x) = 1. \tag{4.6}$$

Then there is a unique moment functional $\mathcal{L}$ such that

$$\mathcal{L}[1] = \lambda_1, \qquad \mathcal{L}[P_m(x)P_n(x)] = 0 \qquad \text{for } m \neq n, \qquad m, n = 0, 1, 2, \ldots.$$

$\mathcal{L}$ is quasi-definite and $\{P_n(x)\}$ is the corresponding monic OPS if and only if $\lambda_n \neq 0$ while $\mathcal{L}$ is positive-definite if and only if c_n is real and $\lambda_n > 0$ $(n \geq 1)$.

Proof   We define the moment functional $\mathcal{L}$ inductively by

$$\mathcal{L}[1] = \mu_0 = \lambda_1, \qquad \mathcal{L}[P_n(x)] = 0, \qquad n = 1, 2, 3, \ldots. \tag{4.7}$$

That is, we define μ_1 by the condition, $\mathcal{L}[P_1(x)] = \mu_1 - c_1\mu_0 = 0$, μ_2 by $\mathcal{L}[P_2(x)] = \mu_2 - (c_1 + c_2)\mu_1 + (\lambda_2 - c_1 c_2)\mu_0 = 0$, etc.

Rewriting (4.6) in the form

$$xP_n(x) = P_{n+1}(x) + c_{n+1}P_n(x) + \lambda_{n+1}P_{n-1}(x), \qquad n \geq 1, \tag{4.8}$$

we obtain because of (4.7)

$$\mathcal{L}[xP_n(x)] = 0, \qquad n \geq 2.$$

Multiplying both sides of (4.8) by x and using the last result, we then find

$$\mathcal{L}[x^2 P_n(x)] = 0, \qquad n \geq 3.$$

Continuing in this manner, we conclude

$$\mathcal{L}[x^k P_n(x)] = 0, \qquad 0 \leq k < n,$$

$$\mathcal{L}[x^n P_n(x)] = \lambda_{n+1}\mathcal{L}[x^{n-1} P_{n-1}(x)], \qquad n \geq 1.$$

It follows that for $m \neq n$, $\mathcal{L}[P_m(x)P_n(x)] = 0$, while exactly as in the proof of Theorem 3.2, we find

$$\mathcal{L}[P_n^2(x)] = \mathcal{L}[x^n P_n(x)] = \lambda_1 \lambda_2 \cdots \lambda_{n+1}, \qquad n \geq 0.$$

Therefore $\mathcal{L}$ is quasi-definite and $\{P_n(x)\}$ is the corresponding OPS if and only if $\lambda_n \neq 0$ for $n \geq 1$.

Finally, the moments are clearly real if c_n and λ_n are all real so by Theorem 3.4, $\mathcal{L}$ is positive-definite if and only if $\lambda_n > 0$ for $n \geq 1$.   ∎

It may be noted that $\{P_n(x)\}$ is independent of λ_1 (see the remark at the end of §2).

The original theorem of Favard concerned only the positive-definite case and the functional $\mathcal{L}$ was represented by a Stieltjes integral. The corresponding result for the quasi-definite case was subsequently observed by Shohat.

In the event $\lambda_{N+1} = 0$ for some N, then $\mathcal{L}[P_n^2(x)] = 0$ for $n \geq N$ and $\{P_n(x)\}$ is not an OPS although it still satisfies the orthogonality condition, $\mathcal{L}[P_m(x)P_n(x)] = 0$ $(m \neq n)$. This situation has been called "weak orthogonality" and has received some study but we will not treat it generally in this book.

We next obtain the "Christoffel-Darboux Identity."

THEOREM **4.5**   Let $\{P_n(x)\}$ satisfy (4.6) with $\lambda_n \neq 0$ $(n \geq 1)$. Then

$$\sum_{k=0}^{n} \frac{P_k(x)P_k(u)}{\lambda_1\lambda_2\cdots\lambda_{k+1}} = (\lambda_1\lambda_2\cdots\lambda_{n+1})^{-1}\frac{P_{n+1}(x)P_n(u) - P_n(x)P_{n+1}(u)}{x - u}. \quad (4.9)$$

Proof   From (4.6), we have for $n \geq 0$ the identities

$$xP_n(x)P_n(u) = P_{n+1}(x)P_n(u) + c_{n+1}P_n(x)P_n(u) + \lambda_{n+1}P_{n-1}(x)P_n(u),$$

$$uP_n(u)P_n(x) = P_{n+1}(u)P_n(x) + c_{n+1}P_n(u)P_n(x) + \lambda_{n+1}P_{n-1}(u)P_n(x).$$

Subtracting the second equation from the first yields

$$(x - u)P_n(x)P_n(u) = P_{n+1}(x)P_n(u) - P_{n+1}(u)P_n(x)$$
$$- \lambda_{n+1}[P_n(x)P_{n-1}(u) - P_n(u)P_{n-1}(x)].$$

If we denote the right side of (4.9) by $F_n(x, u)$, the last equation can be rewritten

$$\frac{P_m(x)P_m(u)}{\lambda_1\lambda_2\cdots\lambda_{m+1}} = F_m(x, u) - F_{m-1}(x, u), \qquad m \geq 0.$$

Summing the latter from 0 to n and noticing that $F_{-1}(x, u) = 0$, we obtain (4.9).   ∎

For the corresponding orthonormal polynomials $p_n(x)$, which can be written

$$p_n(x) = k_n P_n(x), \qquad k_n = (\lambda_1\lambda_2\cdots\lambda_{n+1})^{-1/2}, \qquad (4.10)$$

(4.9) takes the form (in which it most often appears in the literature)

$$\sum_{k=0}^{n} p_k(x)p_k(u) = \frac{k_n}{k_{n+1}}\frac{p_{n+1}(x)p_n(u) - p_n(x)p_{n+1}(u)}{x - u}. \qquad (4.11)$$

THEOREM **4.6**   The following "confluent form" of (4.9) is also valid.

$$\sum_{k=0}^{n} \frac{P_k^2(x)}{\lambda_1 \lambda_2 \cdots \lambda_{k+1}} = \frac{P'_{n+1}(x) P_n(x) - P'_n(x) P_{n+1}(x)}{\lambda_1 \lambda_2 \cdots \lambda_{n+1}}. \qquad (4.12)$$

Proof   The numerator of the right side of (4.9) can be written

$$P_{n+1}(x) P_n(u) - P_n(x) P_{n+1}(u) = [P_{n+1}(x) - P_{n+1}(u)] P_n(x)$$
$$- [P_n(x) - P_n(u)] P_{n+1}(x)$$

so (4.12) follows from (4.9) by letting $u \to x$.   ∎

As an immediate corollary, we obtain the important inequality

$$P'_{n+1}(x) P_n(x) - P'_n(x) P_{n+1}(x) > 0, \qquad (4.13)$$

valid for all real x whenever $\mathcal{L}$ is positive-definite.

REMARK   As we have already noted, an OPS defined by (4.6) is independent of λ_1. The convention that $\lambda_1 = \mathcal{L}[1]$ is convenient and will be maintained when we are discussing the general theory. On the other hand, it is frequently convenient to be able to drop this convention when dealing with a specific recurrence formula. Hence it will be understood that when a recurrence formula is given explicitly for a specific OPS, the coefficient of $P_{-1}(x)$ is not necessarily $\mathcal{L}[1]$. Note that otherwise, in the recurrence formula of Ex. 4.10, the case $n = 0$ would have to be written separately.

Exercises

4.1   Referring to (4.1), show that the coefficient of x^{n-2} in $P_n(x)$ is

$$\sum c_i c_j - \sum_{k=2}^{n} \lambda_k$$

where the first sum extends over all i and j such that $1 \leqq i < j \leqq n$.

4.2   Express the coefficients A_n, B_n, and C_n of (4.2) directly in terms of $P_n(x)$ and k_n.

4.3   Show that if $\{P_n(x)\}$ satisfies (4.6) and $\lambda_n \neq 0$ for $n \leqq N$, then $P_n(x)$ and $P_{n-1}(x)$ cannot have a common zero for any $n \leqq N$.

4.4 Let $\{P_n(x)\}$ satisfy (4.6) and let $Q_n(x) = a^{-n}P_n(ax + b)$ $(a \neq 0)$.
   (a) Show that

$$Q_n(x) = \left(x - \frac{c_n - b}{a}\right)Q_{n-1}(x) - \frac{\lambda_n}{a^2}Q_{n-2}(x).$$

   (b) If $\{P_n(x)\}$ is the OPS with respect to the moments μ_n, then $\{Q_n(x)\}$ is an OPS
        with respect to the moments m_n given by

$$m_n = a^{-n}\sum_{k=0}^{n}\binom{n}{k}(-b)^{n-k}\mu_k.$$

4.5 Suppose $\{Q_n(x)\}$ satisfies (4.6) but with the initial conditions $Q_{-1}(x) = -1$ and
     $Q_0(x) = 0$.
   (a) Show that $Q_n(x)$ is a polynomial of degree $n - 1$.
   (b) Put $P_n^{(1)}(x) = \lambda_1^{-1}Q_{n+1}(x)$ and write the recurrence formula satisfied by
        $\{P_n^{(1)}(x)\}$.

4.6 Let $\{P_n(x)\}$ satisfy (4.6) with $c_n = 0$ and $\lambda_n < 0$ for $n \geq 1$. Then $\{P_n(x)\}$ is an OPS
     with respect to a quasi-definite moment functional $\mathcal{L}$. Put $\mathcal{L}^*[x^n] = i^{-n}\mathcal{L}[x^n]$ and
     show that $\mathcal{L}^*$ is positive-definite by obtaining the corresponding monic OPS.

4.7 Suppose $\{P_n(x)\}$ satisfies (4.6) with $c_n = 0$ and $\lambda_n < 0$ for $n \geq 2$ but for $n = 1$,
     $P_1(x) = (x - c_1)P_0(x)$ where $c_1 \neq 0$, $P_0(x) = 1$. Let $R_n(x) = \mathrm{Re}[i^{-n}P_n(ix)]$ and
     $I_n(x) = \mathrm{Im}[i^{-n}P_n(ix)]$ (where $\mathrm{Re}(z)$ and $\mathrm{Im}(z)$ denote the real and imaginary parts
     of a complex number z). Prove that $\{R_n(x)\}_{n=0}^{\infty}$ and $\{c_1^{-1}I_{n+1}(x)\}_{n=0}^{\infty}$ are both
     monic OPS with respect to positive-definite moment functionals.

4.8 (a) If $\{P_n(x)\}$ satisfies (4.2) with $B_n = 0$ for $n \geq 0$, compute $P_{2n}(0)$.
     (b) Show that for the Hermite polynomials (Ex. 1.5): $H_{2n}(0) = (-1)^n(2n)!/n!$;
          for the Legendre polynomials (Ex. 1.6): $P_{2n}(0) = (-1)^n(2n)!/(2^n n!)^2$; for the
          Tchebichef polynomials of the first kind: $T_{2n}(0) = (-1)^n$.

4.9 For the Tchebichef polynomials of the second kind (Ex. 1.2), obtain the
     recurrence formula: $U_n(x) = 2xU_{n-1}(x) - U_{n-2}(x)$ $(n \geq 1)$.

4.10 Let $C_n^{(a)}(x)$ denote the monic Charlier polynomials, $C_n^{(a)}(x) = n!\,P_n(x)$, where
     $P_n(x)$ is defined by (1.13). Obtain the recurrence formula:

$$C_{n+1}^{(a)}(x) = (x - n - a)C_n^{(a)}(x) - anC_{n-1}^{(a)}(x) \;(n \geq 0).$$

4.11 (a) Let $G(x,w) = \sum_{n=0}^{\infty}T_n(x)w^n$. Use the recurrence formula (4.5) to derive an
          algebraic equation satisfied by G and thus obtain

$$G(x,w) = \frac{1 - xw}{1 - 2xw + w^2}.$$

     (b) Similarly derive the generating function,

$$\frac{1}{1 - 2xw + w^2} = \sum_{n=0}^{\infty}U_n(x)w^n.$$

4.12 Prove that if $P_n(x)$ is defined by (4.6), then

$$P_n(x) = \begin{vmatrix} x - c_1 & 1 & 0 & 0 & \cdots & 0 & 0 & 0 & 0 \\ \lambda_2 & x - c_2 & 1 & 0 & \cdots & 0 & 0 & 0 & 0 \\ 0 & \lambda_3 & x - c_3 & 1 & \cdots & 0 & 0 & 0 & 0 \\ . & . & . & . & \cdots & . & . & . & . \\ . & . & . & . & \cdots & . & . & . & . \\ 0 & 0 & 0 & 0 & \cdots & 0 & \lambda_{n-1} & x - c_{n-1} & 1 \\ 0 & 0 & 0 & 0 & \cdots & 0 & 0 & \lambda_n & x - c_n \end{vmatrix}$$

4.13 Prove that an OPS $\{P_n(x)\}$ satisfies a relation of the form

$$P_{n-1}(x)P_n(-x) + P_{n-1}(-x)P_n(x) = a_n \neq 0, \qquad n \geq 1,$$

if and only if in its recurrence formula (4.6), we have $\lambda_n \neq 0$ $(n \geq 1)$ and $c_n = 0$ for $n \geq 2$ but $c_1 \neq 0$. Show further that the condition for positive-definiteness is $(-1)^n a_1 a_n < 0$ for $n \geq 2$. (D. Dickinson [1])

4.14 Let $\{P_n(x)\}$ be an OPS; let $\mathfrak{L}$ be a moment functional such that $\mathfrak{L}[1] \neq 0$, $\mathfrak{L}[P_n(x)] = 0$ for $n > 0$. Prove that $\{P_n(x)\}$ is an OPS with respect to $\mathfrak{L}$.

4.15 Let $\{P_n(x)\}$ satisfy (4.1) and let $\{Q_n(x)\}$ satisfy

$$Q_n(x) = (x - d_n)Q_{n-1}(x) - \nu_n Q_{n-2}(x).$$

Let also $Q_n(x) = \sum_{k=0}^{n} a_{nk} P_k(x)$. Prove that if for all k and m, $1 \leq k \leq m \leq N$, we have $c_k \geq d_m$ and $\lambda_k \geq \nu_m$, then $a_{mk} \geq 0$ for $0 \leq k \leq m \leq N$. (Askey [8])

5 Zeros

When the moment functional is positive-definite, the zeros of the corresponding orthogonal polynomials exhibit a certain regularity in their behavior. In order to discuss this behavior, we will have to make a refinement in our concept of positive-definiteness.

DEFINITION 5.1 Let $E \subset (-\infty, \infty)$. A moment functional $\mathfrak{L}$ is said to be *positive-definite on* E if and only if $\mathfrak{L}[\pi(x)] > 0$ for every real polynomial $\pi(x)$ which is non-negative on E and does not vanish identically on E. The set E is called a *supporting set* for $\mathfrak{L}$.

For example, the moment functionals for the Tchebichef polynomials of both the first and second kinds are positive-definite on $(-1, 1)$ while the moment functionals for the Charlier and Hermite polynomials (see the Exercises, §1) are positive-definite on $\{0, 1, 2, \ldots\}$ and $(-\infty, \infty)$, respectively. Of course, positive-definiteness on $(-\infty, \infty)$ is the same thing as positive-definiteness. The next theorem shows that positive-definiteness on any infinite set implies positive-definiteness (but not conversely).

THEOREM **5.1** If $\mathcal{L}$ is positive-definite on E and E is an infinite set, then (i) $\mathcal{L}$ is positive-definite on every set containing E and (ii) $\mathcal{L}$ is positive-definite on every dense subset of E.

Proof Let $\pi(x)$ be a real polynomial which is non-negative and does not vanish identically on a set S.

If $S \supset E$, then trivially, $\pi(x) \geqq 0$ on E. If, on the other hand, $S \subset E$ but S is dense in E, then $\pi(x) \geqq 0$ by continuity. Since $\pi(x)$ cannot vanish everywhere on an infinite set, it follows in either case that $\mathcal{L}[\pi(x)] > 0$. ∎

The conclusion (i) of Theorem 5.1 does not hold if E is a finite set (see Ex. 5.1).

Theorem 5.1 (ii) shows that in general there is no "smallest" infinite supporting set for a positive-definite moment functional. However, it is natural to ask whether there is a smallest *closed* supporting set F for $\mathcal{L}$ in the sense that F is a subset of every closed supporting set for $\mathcal{L}$.

The answer is in the affirmative if $\mathcal{L}$ has a *bounded* supporting set but not necessarily so in general. If all supporting sets are unbounded, the question of a smallest closed supporting set is equivalent to the question of the determinacy of the corresponding Hamburger moment problem which will be discussed in Chapter II.

Throughout the remainder of this section, $\mathcal{L}$ will denote a positive-definite moment functional and $\{P_n(x)\}$ the corresponding monic OPS.

THEOREM **5.2** Let I be an interval which is a supporting set for $\mathcal{L}$. The zeros of $P_n(x)$ are all real, simple and are located in the interior of I.

Proof Since $\mathcal{L}[P_n(x)] = 0$, $P_n(x)$ must change sign at least once in the interior of I. That is, $P_n(x)$ has at least one zero of odd multiplicity located in the interior of I.

Let $x_1, x_2, \ldots, x_k$ denote the distinct zeros of odd multiplicity that are located in the interior of I. Set

$$\rho(x) = (x - x_1) \cdots (x - x_k).$$

Then $\rho(x) P_n(x)$ is a polynomial that has no zeros of odd multiplicity in the interior of I, hence $\rho(x) P_n(x) \geqq 0$ for $x \in I$. Therefore, $\mathcal{L}[\rho(x) P_n(x)] > 0$. But this contradicts Theorem 1.1 unless $k \geqq n$. That is, $k = n$ so $P_n(x)$ has n distinct zeros in the interior of I. ∎

Henceforth, we denote the zeros of $P_n(x)$ by x_{ni}, the zeros being ordered by increasing size:

$$x_{n1} < x_{n2} < \cdots < x_{nn}, \qquad n \geqq 1. \tag{5.1}$$

Since $P_n(x)$ has positive leading coefficient, it follows that

$$P_n(x) > 0 \quad \text{for } x > x_{nn}; \qquad \text{sgn } P_n(x) = (-1)^n \quad \text{for } x < x_{n1}. \tag{5.2}$$

Here sgn denotes the *signum function* defined by

$$\text{sgn } x = \begin{cases} 1 & x > 0 \\ 0 & x = 0 \\ -1 & x < 0. \end{cases}$$

Now $P_n'(x)$ has at least one, hence exactly one, zero on each of the intervals, $(x_{n,k-1}, x_{n,k})$. It follows that $P_n'(x_{nk})$ alternates in sign as k varies from 1 through n. Since $P_n'(x)$ also has positive leading coefficient, we can conclude:

$$\text{sgn } P_n'(x_{nk}) = (-1)^{n-k}, \qquad k = 1, 2, \ldots, n. \tag{5.3}$$

THEOREM 5.3 (Separation theorem for the zeros) The zeros of $P_n(x)$ and $P_{n+1}(x)$ mutually separate each other. That is,

$$x_{n+1,i} < x_{ni} < x_{n+1,i+1}, \qquad i = 1, 2, \ldots, n. \tag{5.4}$$

Proof We have the inequality (4.13):

$$P_{n+1}'(x)P_n(x) - P_n'(x)P_{n+1}(x) > 0.$$

In particular,

$$P_{n+1}'(x_{n+1,k})P_n(x_{n+1,k}) > 0, \qquad k = 1, 2, \ldots, n+1. \tag{5.5}$$

Referring to (5.3), we conclude that sgn $P_n(x_{n+1,k}) = (-1)^{n+1-k}$. Thus $P_n(x)$ has at least one, hence exactly one, zero on each of the n intervals, $(x_{n+1,k}, x_{n+1,k+1})$ $(k = 1, 2, \ldots, n)$. ∎

COROLLARY For each $k \geqq 1$, $\{x_{nk}\}_{n=k}^{\infty}$ is a decreasing sequence and $\{x_{n,n-k+1}\}_{n=k}^{\infty}$ is an increasing sequence. In particular, the limits

$$\xi_i = \lim_{n \to \infty} x_{ni}, \qquad \eta_j = \lim_{n \to \infty} x_{n,n-j+1}, \qquad i, j = 1, 2, 3, \ldots, \quad (5.6)$$

all exist, at least in the extended real number system.

DEFINITION 5.2   The closed interval, $[\xi_1, \eta_1]$, is called the *true interval of orthogonality* of the OPS (of $\mathfrak{L}$).

The true interval of orthogonality is the smallest closed interval that contains all of the zeros of all $P_n(x)$. It can be shown (see Ex. 6.4) that the set of all these zeros is a supporting set for $\mathfrak{L}$. The true interval of orthogonality is therefore the smallest closed interval that is a supporting set for $\mathfrak{L}$. We remark that there is no particular significance to be attached to the fact that the true interval of orthogonality is written here as a closed interval. Indeed, it appears in the literature more frequently written as an open interval. Our choice is dictated by a minor convenience for we will later be able to say that the true interval of orthogonality contains a certain set (called the *spectrum*) whereas otherwise we would have to refer to the closure of the true interval of orthogonality.

Returning to the orthogonal polynomials, we note that the separation property of the zeros leads to the following partial fraction decomposition:

$$\frac{P_n(x)}{P_{n+1}(x)} = \sum_{k=1}^{n+1} \frac{a_{nk}}{x - x_k}, \qquad x_k = x_{n+1,k}, \qquad (5.7)$$

where

$$a_{nk} = \frac{P_n(x_k)}{P'_{n+1}(x_k)} > 0.$$

The positivity of the coefficients a_{nk} follows from (5.5).

Finally, we obtain a simple inequality involving (5.7). If x_0 is real and not a zero of $P_{n+1}(x)$, then for any z, real or complex and not a zero of $P_{n+1}(x)$,

$$\left| \frac{P_n(z)}{P_{n+1}(z)} \right| \leq \sum_{k=1}^{n+1} \left| \frac{x_0 - x_k}{z - x_k} \right| \left| \frac{a_{nk}}{x_0 - x_k} \right| \leq K(z) \sum_{k=1}^{n+1} \frac{a_{nk}}{|x_0 - x_k|},$$

where

$$K(z) = \max_{1 \leq k \leq n+1} \left| \frac{x_0 - x_k}{z - x_k} \right|. \qquad (5.8)$$

In particular, for $x_0 < x_1$ or $x_0 > x_{n+1}$, this yields

$$\left| \frac{P_n(z)}{P_{n+1}(z)} \right| \leq K(z) \left| \frac{P_n(x_0)}{P_{n+1}(x_0)} \right|. \qquad (5.9)$$

Exercises

5.1 Let $E = \{t_1, t_2, \ldots, t_N\}$ be any set of real numbers, $t_i \neq t_j$ for $i \neq j$. Let $h_i > 0$ $(1 \leq i \leq N)$. Define $\mathcal{L}$ by

$$\mathcal{L}[x^n] = \sum_{i=1}^{N} h_i t_i^n, \qquad n = 0, 1, 2, \ldots.$$

Prove that $\mathcal{L}$ is positive-definite on E but not on any set of which E is a proper subset.

5.2 Determine explicitly the zeros of the Tchebichef polynomials (both kinds) and note the separation properties of each set.

5.3 Referring to (4.6), show that $\xi_1 < c_n < \eta_1$, where $[\xi_1, \eta_1]$ is the true interval of orthogonality of the corresponding OPS.

5.4 With the notation of this section, let $Q_n(x) = P_n(x) + cP_{n-1}(x)$. Prove that $Q_n(x)$ has n real zeros, y_{nk}, satisfying:

if $c > 0$: $x_{n-1,k-1} < y_{n,k} < x_{nk}$, $k = 1, 2, \ldots, n$; $x_{n-1,0} = -\infty$

if $c < 0$: $x_{nk} < y_{nk} < x_{n-1,k}$, $k = 1, 2, \ldots, n$; $x_{n-1,n} = +\infty$.

5.5 Given real numbers $x_1 < x_2 < \cdots < x_n$, show that for each set of $n - 1$ numbers y_i such that $x_{k-1} < y_{k-1} < x_k$ $(2 \leq k \leq n)$, there is an OPS $\{P_n(x)\}$ with respect to a positive-definite moment functional, such that

$$P_{n-1}(x) = (x - y_1) \cdots (x - y_{n-1})$$

and $P_n(x) = (x - x_1) \cdots (x - x_n)$. (B. Wendroff [1]).

5.6 Using the notation of this section, assume that $\xi_1 = 0$ so the zeros of $P_n(x)$ are all positive. Then the *reversed polynomial* $R_n(x) = x^n P_n(1/x)$ has n positive distinct zeros $y_{nk} = x_{n,n-k+1}^{-1}$, and these zeros enjoy the separation property (5.4). Show that, however, $\{R_n(x)\}$ is not an OPS.

5.7 Let $\{P_n(x)\}$ satisfy (4.6) with c_n real and $\lambda_n > 0$ $(n \geq 1)$. Write $\lambda_n = b_n \bar{b}_n$ where $b_n \neq 0$ and consider the Hermitian matrix

$$M_n = \begin{bmatrix} c_1 & b_2 & 0 & 0 & \cdots & 0 & 0 & 0 \\ \bar{b}_2 & c_2 & b_3 & 0 & \cdots & 0 & 0 & 0 \\ 0 & \bar{b}_3 & c_3 & b_4 & \cdots & 0 & 0 & 0 \\ \cdot & \cdot & \cdot & \cdot & \cdots & \cdot & \cdot & \cdot \\ 0 & 0 & 0 & 0 & \cdots & \bar{b}_{n-1} & c_{n-1} & b_n \\ 0 & 0 & 0 & 0 & \cdots & 0 & \bar{b}_n & c_n \end{bmatrix}$$

Prove that the zeros of $P_n(x)$ are the eigenvalues of M_n.

5.8 Let $-\infty < \xi_1 < \xi_2 < \xi_3 < \cdots < \xi_k$. Prove that there is an $N = N(k)$ such that $P_n(x)$ has exactly one zero on each open interval (ξ_i, ξ_{i+1}) for $1 \leqq i < k$ for all $n \geqq N$.

6 Gauss Quadrature

Let $\{t_1, t_2, \ldots, t_n\}$ be any set of $n \geqq 1$ distinct numbers and set

$$F(x) = \prod_{i=1}^{n} (x - t_i).$$

Then $F(x)/(x - t_i)$ is a polynomial of degree $n - 1$ and

$$\lim_{x \to t_k} \frac{F(x)}{x - t_k} = F'(t_k) \neq 0.$$

It follows that

$$l_k(x) = \frac{F(x)}{(x - t_k)F'(t_k)}$$

is a polynomial of degree $n - 1$ with the property

$$l_k(t_j) = \delta_{jk}.$$

Now for any set $\{y_1, y_2, \ldots, y_n\}$ of numbers, the polynomial

$$L_n(x) = \sum_{k=1}^{n} y_k l_k(x)$$

is of degree, at most, $n - 1$ and has the property

$$L_n(t_j) = \sum_{k=1}^{n} y_k \delta_{jk} = y_j, \qquad j = 1, 2, \ldots, n.$$

$L_n(x)$ is the *Lagrange interpolation polynomial* corresponding to the *nodes* (abscissas) t_i and the *ordinates* y_i, and provides a solution to the problem of constructing a polynomial of degree at most $n - 1$ whose graph passes through the points (t_i, y_i) $(i = 1, 2, \ldots, n)$. ($L_n(x)$ is the unique solution. (Ex. 6.1)).

The Lagrange interpolation polynomial will now be used to obtain the renowned *Gauss quadrature formula*. This formula is of considerable utility in numerical analysis for the approximation of integrals. However, our primary interest in it is as an important tool for studying the moment functional $\mathfrak{L}$ in the positive-definite case. In particular, it will play a key role in the representation theorem of Chapter 2.

We assume that $\mathfrak{L}$ is positive-definite and maintain the notation of the preceding section.

THEOREM **6.1** (Gauss quadrature formula). Let $\mathfrak{L}$ be positive-definite. There are numbers $A_{n1}, A_{n2}, \ldots, A_{nn}$ such that for every polynomial $\pi(x)$ of degree at most $2n - 1$,

$$\mathfrak{L}[\pi(x)] = \sum_{k=1}^{n} A_{nk}\pi(x_{nk}). \tag{6.1}$$

The numbers A_{nk} are all positive and satisfy the condition,

$$A_{n1} + A_{n2} + \cdots + A_{nn} = \mu_0. \tag{6.2}$$

Proof. Let $\pi(x)$ be an arbitrary polynomial whose degree does not exceed $2n-1$, and construct the Lagrange interpolation polynomial which corresponds to the nodes x_{nk} and the ordinates $\pi(x_{nk})(1 \leqq k \leqq n)$. Thus consider

$$L_n(x) = \sum_{k=1}^{n} \pi(x_{nk})l_k(x)$$

where

$$l_k(x) = \frac{P_n(x)}{(x - x_{nk})P_n'(x_{nk})}.$$

Now $Q(x) = \pi(x) - L_n(x)$ is a polynomial of degree at most $2n-1$ which vanishes at $x_{nk} (k = 1, \ldots, n)$. That is,

$$Q(x) = R(x)P_n(x)$$

where $R(x)$ is a polynomial of degree at most $n-1$. By Theorem 2.1,

$$\mathfrak{L}[\pi(x)] = \mathfrak{L}[L_n(x)] + \mathfrak{L}[R(x)P_n(x)] = \mathfrak{L}[L_n(x)]$$

$$= \sum_{k=1}^{n} \pi(x_{nk})\mathfrak{L}[l_k(x)].$$

This yields (6.1) with $A_{nk} = \mathfrak{L}[l_k(x)]$.

If the particular choice $\pi(x) = l_m^2(x)$ is made in (6.1), the result is

$$0 < \mathcal{L}[l_m^2(x)] = \sum_{k=1}^n A_{nk} l_m^2(x_{nk}) = A_{nm}$$

so the A_{nk} are all positive.

Finally, (6.2) can be obtained by choosing $\pi(x) = 1$ in (6.1).   ■

It should be noted that the *weights* A_{nk} in the Gauss quadrature formula do not depend on the fact that $P_n(x)$ is monic.

In the event $\mathcal{L}$ is defined as in (1.7), we can consider

$$\mathcal{L}[f] = \int_a^b f(x)w(x)dx$$

provided the integral converges. In this case, (6.1) suggests the approximation

$$\mathcal{L}[f] = \int_a^b f(x)w(x)dx \approx \sum_{k=1}^n A_{nk} f(x_{nk}) \equiv \mathcal{L}_n[f]. \tag{6.3}$$

Formulas of the general form (6.3) in which A_{nk} and x_{nk} are numbers independent of f are called *approximate quadrature* (or *mechanical quadrature*) formulas. If the formula is *exact* (that is, $\mathcal{L}[f] = \mathcal{L}_n[f]$) whenever f is a polynomial of degree m, the quadrature formula is said to have a *degree of precision m*. The Gauss quadrature formula is thus an approximate quadrature formula in which the nodes x_{nk} are the zeros of the appropriate orthogonal polynomials and has degree of precision $2n - 1$. It is in fact the only quadrature formula (with n nodes) with degree of precision $2n - 1$ (see Ex. 6.7).

When f is not a polynomial, questions of convergence and closeness of approximation in (6.3) are naturally of fundamental importance. Not surprisingly, such questions are also generally difficult and delicate. We will not attempt to go into these questions here beyond making the following remarks concerning convergence when the true interval of orthogonality of $\mathcal{L}$ is bounded.

If $[a, b] = [\xi_1, \eta_1]$ is compact and f is continuous on $[a, b]$, then by the Weierstrass approximation theorem, f can be uniformly approximated by polynomials on $[a, b]$. Thus for arbitrary $\epsilon > 0$, there is a polynomial of degree m such that

$$|\mathcal{L}_n[f] - \mathcal{L}_n[\pi(x)]| < \epsilon, \qquad |\mathcal{L}[f] - \mathcal{L}[\pi(x)]| < \epsilon.$$

Thus if $\mathcal{L}_n[f]$ denotes the Gauss quadrature formula with $2n - 1 \geqq m$, then $\mathcal{L}_n[\pi(x)] = \mathcal{L}[\pi(x)]$ and it follows that $\mathcal{L}_n[f] \to \mathcal{L}[f]$ as $n \to \infty$. The routine details are left to the reader.

Theorem 6.1 can be used to provide additional information about the separation properties of the zeros of orthogonal polynomials.

THEOREM 6.2 Between any two zeros of $P_N(x)$ there is at least one zero of $P_n(x)$ for every $n > N \geqq 2$.

Proof Assume that for some $n > N$, $P_n(x)$ has no zero between x_{Np} and $x_{N,p+1}$ $(1 \leqq p < N)$. Now

$$\rho(x) = \frac{P_N(x)}{(x - x_{Np})(x - x_{N,p+1})}$$

is a polynomial of degree $N - 2$ and

$$\rho(x)P_N(x) \geqq 0 \qquad \text{for } x \notin (x_{Np}, x_{N,p+1}). \tag{6.4}$$

Using (6.1) we have

$$\mathcal{L}[\rho(x)P_N(x)] = \sum_{n=1}^{n} A_{nk}\rho(x_{nk})P_N(x_{nk}).$$

Since $\rho(x)P_N(x)$ cannot vanish at every x_{nk}, it follows from (6.4) that $\mathcal{L}[\rho(x)P_N(x)] > 0$ and this contradicts the orthogonality properties. ∎

Exercises

6.1 Prove that $Q(x)$ is a polynomial which interpolates the n points (t_i, y_i) $(1 \leqq i \leqq n)$ if and only if $Q(x) = L_n(x) + \pi(x)F(x)$, where $L_n(x)$ is the corresponding Lagrange interpolation polynomial, $\pi(x)$ is a polynomial and

$$F(x) = (x - t_1) \cdots (x - t_n).$$

6.2 Obtain the following formulas for the weights A_{nk} in (6.1).
   (a)
   $$A_{nk} = -\frac{\lambda_1 \lambda_2 \cdots \lambda_{n+1}}{P_{n+1}(x_{nk})P_n'(x_{nk})}$$
   (b) $A_{nk} = \{\sum_{i=0}^{n} p_i^2(x_{nk})\}^{-1}$ where $p_n(x)$ is given by (4.10).

6.3 Show that for the Tchebichef polynomials of the first kind, A_{nk} in the corresponding Gauss quadrature formula is independent of k. (It was shown by K. Posse that this is the only Gauss quadrature formula in which A_{nk} has this property. See Natanson [1, p. 480-481].)

6.4 Show that if $\mathcal{L}$ is positive-definite, then the set of all zeros x_{nk}, $1 \leq k \leq n$, $n = 1, 2, 3, \ldots$, is a supporting set for $\mathcal{L}$.

6.5 Let $\mathcal{L}$ be positive-definite on a compact interval I. Let $\|f\| = \sup_{x \in I} |f(x)|$ for every complex-valued function f defined on I. Prove that $|\mathcal{L}[\pi(x)]| \leq \mu_0 \|\pi\|$ for every polynomial $\pi(x)$.

6.6 Show that (6.1) is not valid generally for polynomials of degree $2n$.

6.7 Let $\mathcal{L}$ be positive-definite and suppose there are positive numbers $B_{n1}, \ldots, B_{nn}$ and real numbers $\zeta_1 < \zeta_2 < \cdots < \zeta_n$ such that

$$\mathcal{L}[\pi(x)] = \sum_{i=1}^{n} B_{nk}\pi(\zeta_i)$$

for every polynomial of degree at most $2n - 1$. Prove that $\zeta_i = x_{ni}$ and $B_{ni} = A_{ni}$ $(i = 1, \ldots, n)$.

6.8 Complete the details of the proof that when the true interval of orthogonality is bounded, the Gauss quadrature formula converges to $\mathcal{L}[f]$ for every continuous function f.

6.9 Prove that if $\mathcal{L}$ is symmetric, then the weights of the Gauss quadrature formula satisfy $A_{n,n-k+1} = A_{nk}$.

7 Kernel Polynomials

Throughout the remainder of this chapter, we assume that $\mathcal{L}$ is quasi-definite, $\{\mu_n\}$ is its moment sequence and $\{P_n(x)\}$ is the corresponding monic OPS. Whenever $\mathcal{L}$ is further assumed to be positive-definite, the notation introduced in §5 and §6 will be maintained.

For any real or complex number κ, let a new moment functional $\mathcal{L}_\kappa^*$ be defined by

$$\mathcal{L}_\kappa^*[x^n] = \mu_{n+1} - \kappa\mu_n, \qquad n = 0, 1, 2, \ldots. \tag{7.1}$$

It follows immediately that for every polynomial $\pi(x)$

$$\mathcal{L}_\kappa^*[\pi(x)] = \mathcal{L}[(x - \kappa)\pi(x)]. \tag{7.2}$$

Next define polynomials $P_n^*(\kappa; x)$ by

$$P_n^*(\kappa; x) = (x - \kappa)^{-1}\left[P_{n+1}(x) - \frac{P_{n+1}(\kappa)}{P_n(\kappa)}P_n(x)\right], \tag{7.3}$$

where now it must be assumed that κ is not a zero of $P_n(x)$.

THEOREM 7.1 If κ is not a zero of $P_n(x)$ for any n, then $\mathcal{L}_\kappa^*$ is quasi-definite and $\{P_n^*(\kappa; x)\}$ is the corresponding monic OPS.

Moreover, if $\mathcal{L}$ is positive-definite, then $\mathcal{L}_\kappa^*$ is also positive-definite on $[\xi_1, \eta_1]$ if and only if $\kappa \le \xi_1$.

Proof Using (7.2) and (7.3) and recalling Theorem 3.2, we obtain

$$\mathcal{L}_\kappa^*[x^k P_n^*(\kappa; x)] = \mathcal{L}[x^k P_{n+1}(x)] - \frac{P_{n+1}(\kappa)}{P_n(\kappa)}\mathcal{L}[x^k P_n(x)]$$

$$= -\frac{P_{n+1}(\kappa)}{P_n(\kappa)}\mathcal{L}[P_n^2(x)]\delta_{kn}, \qquad 0 \le k \le n.$$

Thus $\{P_n^*(\kappa; x)\}$ is the monic OPS for $\mathcal{L}_\kappa^*$.

If $\mathcal{L}$ is positive-definite and if $-\infty < \kappa \le \xi_1$, it follows at once from (7.2) that $\mathcal{L}_\kappa^*$ is positive-definite on $[\xi_1, \eta_1]$.

Conversely, suppose $\mathcal{L}_\kappa^*$ is positive-definite on $[\xi_1, \eta_1]$. Put

$$\rho(x) = (x - x_{n1})^{-1} P_n(x).$$

Application of the Gauss quadrature formula together with (7.2) yields

$$0 < \mathcal{L}_\kappa^*[\rho^2(x)] = \mathcal{L}[(x - \kappa)\rho^2(x)] = A_{n1}(x_{n1} - \kappa)\rho^2(x_{n1}).$$

This shows that $\kappa < x_{n1}$ and hence that $\kappa \le \xi_1$. ∎

When $\mathcal{L}$ is positive-definite and $-\infty < \kappa \le \xi_1$, $P_n^*(\kappa; x)$ has real and simple zeros $x_{nk}^* = x_{nk}^*(\kappa)$ for which there is the following separation theorem.

THEOREM 7.2 Let $\mathcal{L}$ be positive-definite and $-\infty < \kappa \le \xi_1$. Then

$$x_{nk} < x_{nk}^* < x_{n+1,k+1}, \qquad k = 1, 2, \ldots, n. \tag{7.4}$$

Thus the true interval of orthogonality of $\mathcal{L}_\kappa^*$ is of the form $[\xi_1^*, \eta_1]$ where $\xi_1 \le \xi_1^* \equiv \xi_1^*(\kappa)$.

Proof Using (7.3) it is easily concluded that $P_n^*(\kappa; x_{nk})$ alternates in sign as k runs from 1 through n while $P_n^*(\kappa; x_{n+1,k})$ also alternates in sign as k runs from 1 through $n + 1$. It is further seen that

$$\mathrm{sgn}[P_n^*(\kappa; x_{nn})] = -1, \qquad \mathrm{sgn}[P_n^*(\kappa; x_{n+1,n+1})] = 1.$$

Because of (5.4), (7.4) now follows. ∎

When $\mathfrak{L}$ is defined by an integral as in (1.7), $\mathfrak{L}_\kappa^*$ can be represented by

$$\mathfrak{L}_\kappa^*[x^n] = \int_a^b x^n(x - \kappa)w(x)\,dx.$$

For example, corresponding to the Tchebichef polynomials, $T_n(x)$, we have $P_n(x) = 2^{1-n}\,T_n(x) = 2^{1-n}\cos n\theta$ ($x = \cos\theta$) so that (7.3) becomes

$$P_n^*(\kappa; x) = 2^{-n}(x - \kappa)^{-1}\left[\cos(n + 1)\theta - \frac{T_{n+1}(\kappa)}{T_n(\kappa)}\cos n\theta\right] \qquad (7.5)$$

and $\{P_n^*(\kappa; x)\}$ is the monic OPS with respect to the non-negative weight function $(x - \kappa)(1 - x^2)^{-1/2}$ on $[-1, 1]$ for $\kappa \le -1$.

When $\kappa = -1$, the weight function becomes $(1 - x)^{-1/2}(1 + x)^{1/2}$ and the corresponding polynomials take the appealing form

$$2^n P_n^*(-1; x) \equiv V_n(x) = \frac{\cos(n + \frac{1}{2})\theta}{\cos\dfrac{\theta}{2}}. \qquad (7.6)$$

Returning to the general case, compare (7.3) with the Christoffel-Darboux identity (4.9). We see that

$$P_n^*(\kappa; x) = \lambda_1\lambda_2\cdots\lambda_{n+1}[P_n(\kappa)]^{-1}\sum_{k=0}^n p_k(\kappa)p_k(x) \qquad (7.7)$$

where $p_k(x)$ denotes the orthonormal polynomial given by (4.12).

In the literature, the polynomials

$$K_n(z, x) = \sum_{m=0}^n \overline{p_m(z)}p_m(x) \qquad (7.8)$$

are usually called *Kernel polynomials*.

If $\mathfrak{L}$ has real moments so that $P_n(x)$ is real, then $\overline{p_m(\kappa)} = p_m(\bar{\kappa})$ and we see that $P_n^*(\kappa; x)$ is the monic form of $K_n(\bar{\kappa}, x)$. In particular, when κ is real (which is our most important case) we would have

$$P_n^*(\kappa; x) = \lambda_1\lambda_2\cdots\lambda_{n+1}[P_n(\kappa)]^{-1}K_n(\kappa, x).$$

Therefore, even though it will involve a slight misnomer when complex moments or non-real κ are involved, we will refer to $P_n^*(\kappa; x)$ as the *monic kernel polynomials* corresponding to $\mathfrak{L}$ (or to $\{P_n(x)\}$) and the *K-parameter* κ. We will also refer to $\{P_n^*(\kappa; x)\}$ as the corresponding monic KPS. No confusion should arise over this dual usage of the term, "kernel polynomial."

The ordinary kernel polynomials, $K_n(z,x)$, have a number of interesting and significant properties. One for which we will have later use is the following extremal property.

THEOREM 7.3 Let $\mathcal{L}$ be positive-definite and let z_0 be an arbitrary complex number. Then

$$[K_n(z_0,z_0)]^{-1} = \min \mathcal{L}[|\pi(x)|^2],$$

where the minimum is computed as $\pi(x)$ ranges over all complex polynomials of degree at most n subject to the condition $\pi(z_0) = 1$. The minimum is attained for $\pi(x) = [K_n(z_0,z_0)]^{-1}K_n(z_0,x)$.

Proof Consider any polynomial (cf. Theorem 2.2)

$$\pi(x) = \sum_{k=0}^{n} c_k p_k(x), \qquad c_k = \mathcal{L}[\pi(x)p_k(x)],$$

such that

$$\pi(z_0) = \sum_{k=0}^{n} c_k p_k(z_0) = 1. \tag{7.9}$$

Recalling that for positive-definite $\mathcal{L}$,

$$\mathcal{L}[\overline{p_j(x)}p_k(x)] = \mathcal{L}[p_j(x)p_k(x)] = \delta_{jk},$$

we obtain

$$\mathcal{L}[|\pi(x)|^2] = \mathcal{L}[\overline{\pi(x)}\pi(x)] = \sum_{j,k=0}^{n} \bar{c}_j c_k \delta_{jk} = \sum_{k=0}^{n} |c_k|^2.$$

Applying Cauchy's inequality to (7.9), we obtain

$$1 \le \sum_{k=0}^{n} |c_k|^2 \sum_{k=0}^{n} |p_k(z_0)|^2 = \mathcal{L}[|\pi(x)|^2]K_n(z_0,z_0),$$

where equality holds if and only if

$$c_k = A\overline{p_k(z_0)}, \qquad A = [K_n(z_0,z_0)]^{-1}.$$

That is,

$$\mathcal{L}[|\pi(x)|^2] \ge [K_n(z_0,z_0)]^{-1}$$

with equality if and only if $\pi(x) = A \sum_{k=0}^{n} \overline{p_k(z_0)}p_k(x)$. ∎

Exercises

7.1 Let z_0 be an arbitrary complex number. Determine the maximum of $|\pi(z_0)|^2$ as $\pi(x)$ ranges over all polynomials of degree at most n such that $\mathcal{L}[|\pi(x)|^2] = 1$.

7.2 For a fixed κ, a monic OPS $\{P_n(x)\}$ uniquely determines $\{P_n^*(\kappa; x)\}$. Show that, however, there are infinitely many different monic OPS whose monic KPS with K-parameter κ is $\{P_n^*(\kappa; x)\}$.

7.3 If $\mathcal{L}$ is positive-definite, $\eta_1 \leq \kappa < \infty$, then define $\mathcal{L}^*$ by $\mathcal{L}^*[x^n] = \mathcal{L}[(\kappa - x)x^n]$. Show that $\mathcal{L}^*$ is positive-definite on $[\xi_1, \eta_1]$ and determine the corresponding monic OPS.

7.4 Verify (7.6).

7.5 Obtain the analogue of (7.5) for the Tchebichef polynomials of the second kind, $\{U_n(x)\}$. In particular, obtain

$$U_n^*(-1; x) = \frac{(n + 1)\sin(n + 2)\theta + (n + 2)\sin(n + 1)\theta}{(n + 1)2^{n+1}(1 + \cos \theta)\sin \theta}, \qquad x = \cos \theta,$$

as the monic orthogonal polynomials with respect to the weight function, $(1 - x)^{1/2}(1 + x)^{3/2}$ on $(-1, 1)$.

7.6 Show that

$$W_n(x) = \frac{\sin[(2n + 1)(\theta/2)]}{\sin(\theta/2)}, \qquad x = \cos \theta,$$

are the orthogonal polynomials with leading coefficients 2^n with respect to the weight function $(1 - x)^{1/2}(1 + x)^{-1/2}$ on $(-1, 1)$.

7.7 If $P_n(x)$ denotes the Legendre polynomial (Ex. 1.6), show that

$$P_{n+1}(x) + P_n(x) = \frac{1 \cdot 3 \cdots (2n + 1)}{(n + 1)!}(x + 1)Q_n(x),$$

where $\{Q_n(x)\}$ is the monic OPS with respect to the weight function $(1 + x)$ on $(-1, 1)$.

7.8 The *Jacobi polynomials*, $P_n^{(\alpha,\beta)}(x)$, are defined as the orthogonal polynomials with respect to the weight function $(1 - x)^\alpha(1 + x)^\beta (\alpha > -1, \beta > -1)$ on $(-1, 1)$, standardized by choosing the leading coefficient to be

$$k_n^{(\alpha,\beta)} = 2^{-n}\binom{\alpha + \beta + 2n}{n}$$

$$= \frac{(\alpha + \beta + n + 1)(\alpha + \beta + n + 2)\cdots(\alpha + \beta + 2n)}{2^n n!}.$$

Prove that

$$(2n + \alpha + \beta + 2)(1 + x)P_n^{(\alpha,\beta+1)}(x) = 2(n + \beta + 1)P_n^{(\alpha,\beta)}(x)$$
$$+ 2(n + 1)P_{n+1}^{(\alpha,\beta)}(x).$$

[Note the special cases, $(\alpha, \beta) = (0,0)$, $(0, 1)$, $(-1/2, -1/2)$, $(1/2, 1/2)$, $(1/2, 3/2)$, $(-1/2, 1/2)$, $(1/2, -1/2)$ which have appeared thus far in the text and exercises.]

7.9 (*Reproducing property* of the kernel polynomials) Let $\mathcal{L}$ be positive-definite. Prove that for any polynomial, $\pi(x)$,

$$\pi(t) = \mathcal{L}[\pi(x)K_n(x, t)].$$

(It is understood that $\mathcal{L}$ operates on x.)

7.10 Suppose $\mathcal{L}$ is defined by an integral as in (1.7) (or more generally as in (3.6)). Let $\{p_n(x)\}$ denote the corresponding orthonormal polynomials. Assuming all integrals exist, let

$$s_n(x) = s_n(f; x) = \sum_{k=0}^{n} c_k p_k(x), \qquad c_k = \mathcal{L}[f(x)p_k(x)].$$

Prove that

$$f(t) - s_n(t) = \int_a^b [f(t) - f(x)]K_n(x, t)w(x)\,dx.$$

(The above forms a basis for investigations into the convergence of expansions of "arbitrary" functions into series of the $p_n(x)$.)

7.11 Let a positive-definite moment functional $\mathcal{L}$ be defined by $\mathcal{L}[x^k] = \int_0^\infty x^k w(x)\,dx$ $< \infty$, where $w(x) \geq 0$ on $[0, \infty)$, and let $\{P_n(x)\}$ be the corresponding monic OPS. For a fixed positive integer ν, let $\{Q_n(x)\}$ denote the monic OPS corresponding to the functional $\mathfrak{M}[x^k] = \int_0^\infty x^k x^\nu w(x)\,dx$. Write $(-1)^n Q_n(x)$ $= \sum_{k=0}^{\infty} (-1)^k \alpha_k P_k(x)$ and prove that $\alpha_k \geq 0$ $(0 \leq k \leq n)$ (Askey [1]). This result remains true if $\nu > 0$ is not an integer (Trench [1]).

7.12 Let $\mathcal{L}$ be positive-definite on an infinite set E and let $\{P_n(x)\}$ be its monic OPS. Let $\rho(x) = (x - x_1) \cdots (x - x_r)$ be non-negative on E, $x_i \neq x_j$ for $i \neq j$.

(a) Show that $D(n, r) = \det(P_{n+i-1}(x_j))_{i,j=1}^r \neq 0$.

(b) Let

$$F_n(x) = \begin{vmatrix} P_n(x_1) & P_{n+1}(x_1) & \cdots & P_{n+r}(x_1) \\ \cdot & \cdot & \cdots & \cdot \\ \cdot & \cdot & \cdots & \cdot \\ P_n(x_r) & P_{n+1}(x_r) & \cdots & P_{n+r}(x_r) \\ P_n(x) & P_{n+1}(x) & \cdots & P_{n+r}(x) \end{vmatrix}$$

and prove that $F_n(x) = \rho(x)Q_n(x)$ where $\{Q_n(x)\}$ is an OPS with respect to the moment functional $\mathfrak{M}$ defined by $\mathfrak{M}[x^k] = \mathcal{L}[x^k \rho(x)]$ (Christoffel's formula).

8 Symmetric Moment Functionals

Let $\mathcal{S}$ be a symmetric quasi-definite moment functional and let $\mathcal{L}$ be defined by

$$\mathcal{L}[x^n] = \mathcal{S}[x^{2n}].$$

If $\{S_n(x)\}$ is the monic OPS for $\mathfrak{S}$, then according to Theorem 4.3, $S_n(-x) = (-1)^n S_n(x)$. It follows that

$$S_{2m}(x) = P_m(x^2), \qquad S_{2m+1}(x) = xQ_m(x^2), \qquad (8.1)$$

where $P_m(x)$ and $Q_m(x)$ are monic polynomials of degree m.

Since we clearly have $\mathfrak{L}[\pi(x)] = \mathfrak{S}[\pi(x^2)]$ for every polynomial $\pi(x)$, it follows that

$$\mathfrak{L}[P_m(x)P_n(x)] = \mathfrak{S}[S_{2m}(x)S_{2n}(x)] \qquad (8.2)$$

$$\mathfrak{L}_0^*[Q_m(x)Q_n(x)] = \mathfrak{L}[xQ_m(x)Q_n(x)] = \mathfrak{S}[S_{2m+1}(x)S_{2n+1}(x)]. \qquad (8.3)$$

This shows that $\{P_n(x)\}$ is the monic OPS for $\mathfrak{L}$ and that

$$Q_n(x) = P_n^*(0; x).$$

Moreover, if $\mathfrak{S}$ is positive-definite, then by the corollary to Theorem 3.4, so are $\mathfrak{L}$ and $\mathfrak{L}_0^*$. Since the zeros of $S_{2m}(x)$ must occur in pairs situated symmetrically about the origin, the true interval of orthogonality of $\mathfrak{S}$ must be of the form $[-\zeta, \zeta]$ and the true interval of orthogonality of $\mathfrak{L}$ must be $[\xi_1, \zeta^2]$ where $\xi_1 \geq 0$.

Conversely, we can begin with a quasi-definite moment functional $\mathfrak{L}$ and define a symmetric moment functional $\mathfrak{S}$ by

$$\mathfrak{S}[x^{2m}] = \mathfrak{L}[x^m], \qquad \mathfrak{S}[x^{2m+1}] = 0, \qquad m \geq 0. \qquad (8.4)$$

If $\{P_n(x)\}$ is the monic OPS for $\mathfrak{L}$ and $Q_n(x) = P_n^*(0; x)$, then $\{S_n(x)\}$ can be defined by (8.1). It follows directly that

$$\mathfrak{S}[S_{2m}(x)S_{2n+1}(x)] = 0, \qquad m, n = 0, 1, 2, \ldots.$$

This together with (8.2) and (8.3) shows that $\{S_n(x)\}$ is the monic OPS for $\mathfrak{S}$.

If in addition $\mathfrak{L}$ is positive-definite and its true interval of orthogonality, $[\xi_1, \eta_1]$, satisfies $\xi_1 \geq 0$, then $\mathfrak{L}_0^*$ is also positive-definite on $[\xi_1, \eta_1]$ (Theorem 7.1). Once again referring to the corollary to Theorem 3.4, we conclude that $\mathfrak{S}$ is positive-definite. The true interval of orthogonality of $\mathfrak{S}$ is readily found to be $[-\eta_1^{1/2}, \eta_1^{1/2}]$.

In summary, we have proven:

THEOREM 8.1 Let $\mathfrak{L}$ and $\mathfrak{S}$ be moment functionals related by (8.4). Let $\{P_n(x)\}$, $\{Q_n(x)\}$ and $\{S_n(x)\}$ be related by (8.1). Then $\{S_n(x)\}$ is the monic OPS for $\mathfrak{S}$ if and only if $\{P_n(x)\}$ is the monic OPS for $\mathfrak{L}$ and $\{Q_n(x)\}$ is the corresponding monic KPS with K-parameter 0.

$\mathcal{S}$ is positive-definite if and only if $\mathcal{L}$ is positive-definite on $[0, \infty]$. In this case, their respective true intervals of orthogonality are $[-\zeta, \zeta]$ and $[\xi_1, \zeta^2]$, where $0 < \zeta \leqq \infty$ and $0 \leqq \xi_1 < \zeta^2$.

If w is an integrable even function, a symmetric moment functional can be defined by

$$\mathcal{S}[x^n] = \int_{-c}^{c} x^n w(x)\, dx$$

and the corresponding functionals, $\mathcal{L}$ and $\mathcal{L}_0^*$ can be represented by

$$\mathcal{L}[x^n] = \int_0^{c^2} x^{n-1/2} w(x^{1/2})\, dx, \qquad \mathcal{L}_0^*[x^n] = \int_0^{c^2} x^{n+1/2} w(x^{1/2})\, dx.$$

(It must be assumed that the integrals converge, of course.)

Again choosing the Tchebichef polynomials $\{T_n(x)\}$ by way of illustration, we obtain $\{T_{2n}(x^{1/2})\}$ and $\{x^{-1/2} T_{2n+1}(x^{1/2})\}$ as the OPS, respectively, corresponding to the weight functions

$$x^{-1/2}(1 - x)^{-1/2} \quad \text{and} \quad x^{1/2}(1 - x)^{-1/2}, \qquad 0 < x < 1.$$

On the other hand, we have

$$(1 - x)^\alpha (1 + x)^\beta = 2^{\alpha+\beta}(1 - u)^\alpha u^\beta, \qquad x = 2u - 1.$$

Thus if $\alpha = \beta = -1/2$, an OPS corresponding to the weight function $(1 - u)^{-1/2} u^{-1/2}$, $0 < u < 1$, is $\{T_n(2u - 1)\}$. Taking leading coefficients into account, we conclude

$$T_{2n}(x^{1/2}) = T_n(2x - 1). \tag{8.5}$$

Similarly, reference to (7.6) shows that $\{V_n(2u - 1)\}$ is an OPS for the weight function $(1 - u)^{-1/2} u^{1/2}$, $0 \leqq u < 1$. This yields the identity

$$T_{2n+1}(x^{1/2}) = x^{1/2} V_n(2x - 1). \tag{8.6}$$

As an example going the other way, consider the *Laguerre polynomials*, $L_n^\alpha(x)$, which are defined as the orthogonal polynomials with leading coefficients $(-1)^n/n!$ corresponding to the weight function

$$v(x) = x^\alpha e^{-x}, \qquad \alpha > -1, \qquad 0 < x < \infty.$$

If $P_n(x) = (-1)^n n! L_n^\alpha(x)$, then the weight function shows that $P_n^*(0; x)$ $= (-1)^n n! L_n^{\alpha+1}(x)$. Therefore the polynomials

$$S_{2n}^\alpha(x) = (-1)^n n! L_n^\alpha(x^2), \qquad S_{2n+1}^\alpha(x) = (-1)^n n! x L_n^{\alpha+1}(x^2)$$

are the monic orthogonal polynomials for the weight function

$$w(x) = |x|^{2\alpha+1} e^{-x^2}, \qquad -\infty < x < \infty.$$

In particular, if $\alpha = -1/2$, this is the weight function for the Hermite polynomials, $H_n(x)$, introduced in Ex. 1.4. For the general case, the polynomials

$$H_n^\mu(x) = 2^n S_n^\alpha(x), \qquad \mu = \alpha + \frac{1}{2}, \tag{8.7}$$

are called *generalized Hermite polynomials*. In particular, we have $H_n^0(x)$ $= H_n(x)$.

Exercises

8.1 Show that the identities (8.5) and (8.6) are elementary trigonometric identities in disguise.

8.2 Derive the identities

$$U_{2n}(x) = W_n(2x^2 - 1), \qquad U_{2n+1}(x) = 2x U_n(2x^2 - 1)$$

in the same manner (8.5) and (8.6) were derived. ($W_n(x)$ is defined in Ex. 7.6.) Verify these are also essentially trigonometric identities.

8.3 Using the notation of this section, show that if

$$\int_0^{c^2} P_m(x) P_n(x) w(x) \, dx = K_n \delta_{mn}, \qquad (c > 0)$$

then

$$\int_{-c}^{c} S_m(x) S_n(x) |x| w(x^2) \, dx = K_n \delta_{mn}.$$

What is the corresponding result if $w(x)dx$ is replaced by a "Stieltjes distribution," $d\psi(x)$?

8.4 Derive the identity involving Laguerre polynomials:

$$x L_n^{\alpha+1}(x) = (n + \alpha + 1) L_n^\alpha(x) - (n + 1) L_{n+1}^\alpha(x).$$

8.5 (a) Prove that

$$(1 - w)^{-\alpha-1} \exp\left\{-\frac{xw}{1 - w}\right\} = \sum_{n=0}^{\infty} L_n^\alpha(x) w^n$$

by using the "method of generating functions" which was illustrated with the Charlier polynomials in § 1 and outlined in Ex. 1.4. to obtain

$$\int_0^\infty L_m^\alpha(x) L_n^\alpha(x) x^\alpha e^{-x}\, dx = \frac{\Gamma(n + \alpha + 1)}{n!} \delta_{mn}.$$

(Γ denotes the *Gamma function*: $\Gamma(a) = \int_0^\infty x^a e^{-x}\, dx$, $a > -1$; $\Gamma(a + 1) = a\Gamma(a)$. See, for example, Rainville [1].)

(b) Use the generating function to derive the recurrence formula,

$$nL_n^\alpha(x) = (-x + 2n + \alpha - 1)L_{n-1}^\alpha(x) - (n + \alpha - 1)L_{n-2}^\alpha(x), \qquad n \geq 1.$$

8.6 Verify that the polynomials $S_n(x)$ defined by

$$S_{2n}(x) = \binom{2n + \alpha + \beta}{n}^{-1} P_n^{(\alpha,\beta)}(2x^2 - 1)$$

$$S_{2n+1}(x) = \binom{2n + \alpha + \beta + 1}{n}^{-1} x P_n^{(\alpha,\beta+1)}(2x^2 - 1) \qquad (\alpha, \beta > -1)$$

are the monic orthogonal polynomials corresponding to the weight function $w(x) = |x|^{2\beta+1}(1 - x^2)^\alpha$, $-1 < x < 1$. (Here $P_n^{(\alpha,\beta)}(x)$ is the Jacobi polynomial defined in Ex. 7.8.)

8.7 The *Gegenbauer (Ultraspherical) polynomials*, $\{P_n^\lambda(x)\}$, are defined for $\lambda > -1/2$ by

$$P_n^\lambda(x) = A_n^\lambda P_n^{(\alpha,\alpha)}(x), \qquad \lambda = \alpha + \frac{1}{2}, \qquad n \geq 0,$$

where

$$A_0^\lambda = 1, \qquad A_n^0 = \frac{2^{n+1}(n - 1)!}{1 \cdot 3 \cdots (2n - 1)}, \qquad n \geq 1,$$

$$A_n^\lambda = \frac{2^n(2\lambda)(2\lambda + 1)\cdots(2\lambda + n - 1)}{(2\lambda + 1)(2\lambda + 3)\cdots(2\lambda + 2n - 1)} \qquad \text{for } \lambda \neq 0, \qquad n \geq 1.$$

$P_n^{(\alpha,\beta)}(x)$ denotes the Jacobi polynomial (Ex. 7.8) so that $\{P_n^\lambda(x)\}$ is an OPS with respect to $(1 - x^2)^{\lambda-1/2}$ on $(-1, 1)$.

(a) Show that the leading coefficient of $P_n^\lambda(x)$ is 1 for $n = 0$ and is given by

$$k_n^0 = \frac{2^n}{n!}, \qquad k_n^\lambda = \frac{2^n \lambda(\lambda + 1)\cdots(\lambda + n - 1)}{n!} \qquad \text{for } n \geq 1.$$

(b) Prove that for $\lambda \neq 0$,

$$P_{2n}^\lambda(x) = \frac{2^n \lambda(\lambda + 1)\cdots(\lambda + n - 1)}{1 \cdot 3 \cdots (2n - 1)} P_n^{(\lambda-1/2,-1/2)}(2x^2 - 1)$$

$$P_{2n+1}^\lambda(x) = \frac{2^{n+1}\lambda(\lambda + 1)\cdots(\lambda + n)}{1 \cdot 3 \cdots (2n + 1)} x P_n^{(\lambda-1/2,1/2)}(2x^2 - 1).$$

Note the special case, $\lambda = 1/2$.
(c) Obtain the corresponding formulas for $\lambda = 0$.

8.8 Let $\{\mu_n^*\}_{n=0}^{\infty}$ be defined by $\mu_{2n}^* = \mu_n$, $\mu_{2n+1}^* = 0$ ($n \geq 0$). Prove that

$$\det(\mu_{i+j}^*)_{i,j=0}^{2n} = [\det(\mu_{i+j})_{i,j=0}^{n}][\det(\mu_{i+j+1})_{i,j=0}^{n-1}].$$

8.9 Obtain separation properties for the zeros of $\{S_n(x)\}$.

9 Certain Related Recurrence Relations

In view of the simple formulas (7.3) and (8.1), it is natural to expect simple relations among the coefficients in the recurrence relations satisfied by the three OPS. To discover these relations, suppose that the recurrences satisfied by $\{P_n(x)\}$, $\{P_n^*(0; x)\}$ and $\{S_n(x)\}$, respectively, are:

$$P_n(x) = (x - c_n)P_{n-1}(x) - \lambda_n P_{n-2}(x), \quad n = 1, 2, 3, \ldots,$$
$$P_{-1}(x) = 0, \quad P_0(x) = 1, \quad \lambda_n \neq 0, \quad \lambda_1 = \mathcal{L}[1]; \tag{9.1}$$

$$Q_n(x) = (x - d_n)Q_{n-1}(x) - \nu_n Q_{n-2}(x), \quad n = 1, 2, 3, \ldots,$$
$$Q_{-1}(x) = 0, \quad Q_0(x) = 1, \quad \nu_n \neq 0, \quad \nu_1 = \mathcal{L}_0^*[1], \tag{9.2}$$

where we have written $Q_n(x) = P_n^*(0; x)$;

$$S_n(x) = xS_{n-1}(x) - \gamma_n S_{n-2}(x), \quad n = 1, 2, 3, \ldots,$$
$$S_{-1}(x) = 0, \quad S_0(x) = 1, \quad \gamma_n \neq 0, \quad \gamma_1 = \mathcal{S}[1], \tag{9.3}$$

where $S_n(x)$ satisfies (8.1). (Note that Theorem 4.3 has been used in writing (9.3).)
For $n = 2m \geq 2$, (9.3) can be written in the form

$$P_m(x^2) = x^2 Q_{m-1}(x^2) - \gamma_{2m} P_{m-1}(x^2).$$

That is,

$$P_m(x) = xQ_{m-1}(x) - \gamma_{2m} P_{m-1}(x), \quad m \geq 1. \tag{9.4}$$

Similarly, (9.3) can be written for $n = 2m + 1 \geq 1$,

$$Q_m(x) = P_m(x) - \gamma_{2m+1} Q_{m-1}(x), \quad m \geq 0. \tag{9.5}$$

We can now use (9.5) to eliminate $P_m(x)$ from (9.4). Then shifting the index in (9.5) from m to $m-1$, we can eliminate $P_{m-1}(x)$ from (9.4). The resulting equation rearranges to

$$Q_m(x) = (x - \gamma_{2m} - \gamma_{2m+1})Q_{m-1}(x) - \gamma_{2m-1}\gamma_{2m}Q_{m-2}(x), \qquad m \geq 1. \tag{9.6}$$

Similarly using (9.4) to eliminate first $Q_m(x)$, then $Q_{m-1}(x)$, from (9.5) leads to the equation,

$$P_{m+1}(x) = (x - \gamma_{2m+1} - \gamma_{2m+2})P_m(x) - \gamma_{2m}\gamma_{2m+1}P_{m-1}(x), \qquad m \geq 1.$$

Referring to (9.4) when $m = 1$, we thus obtain

$$P_m(x) = (x - \gamma_{2m-1} - \gamma_{2m})P_{m-1}(x) - \gamma_{2m-2}\gamma_{2m-1}P_{m-2}(x), \qquad m \geq 2,$$
$$P_1(x) = (x - \gamma_2)P_0(x). \tag{9.7}$$

Comparison of the coefficients in (9.1) and (9.2) with those in (9.6) and (9.7) yields the desired relations:

$$c_1 = \gamma_2, \qquad c_n = \gamma_{2n-1} + \gamma_{2n}, \qquad \lambda_n = \gamma_{2n-2}\gamma_{2n-1}, \qquad n \geq 2;$$
$$d_n = \gamma_{2n} + \gamma_{2n+1}, \qquad n \geq 1, \qquad \nu_n = \gamma_{2n-1}\gamma_{2n}, \qquad n \geq 2.$$

Note also that if we introduce $\gamma_0 = 1$, then

$$\lambda_1 = \mathcal{L}[1] = \mathcal{S}[1] = \gamma_0\gamma_1;$$
$$\nu_1 = \mathcal{L}_0^*[1] = \mathcal{L}[x] = \mathcal{L}[P_1(x) + \gamma_2] = \gamma_1\gamma_2.$$

Since each recurrence formula uniquely determines and is determined by the corresponding moment functional, we have proved:

THEOREM 9.1 Let $\{P_n(x)\}$, $\{Q_n(x)\}$ and $\{S_n(x)\}$ be OPS satisfying (9.1), (9.2) and (9.3), respectively. Then in order that

$$Q_n(x) = P_n^*(0; x), \qquad S_{2m}(x) = P_m(x^2), \qquad S_{2m+1}(x) = xQ_m(x^2), \tag{9.8}$$

it is necessary and sufficient that

$$c_1 = \gamma_2, \qquad c_n = \gamma_{2n-1} + \gamma_{2n}, \qquad n \geq 2, \qquad \lambda_n = \gamma_{2n-2}\gamma_{2n-1}, \tag{9.9}$$
$$n \geq 1;$$
$$d_n = \gamma_{2n} + \gamma_{2n+1}, \qquad \nu_n = \gamma_{2n-1}\gamma_{2n}, \qquad n \geq 1 \qquad (\gamma_0 = 1). \tag{9.10}$$

Moreover all three corresponding moment functionals are positive-definite if and only if $\gamma_n > 0$ $(n \geqq 1)$.

COROLLARY   Let $\{R_n(x)\}$ satisfy the recurrence formula,

$$R_n(x) = (x - f_n)R_{n-1}(x) - \rho_n R_{n-2}(x), \qquad n = 1, 2, 3, \ldots,$$
$$R_{-1}(x) = 0, \qquad R_0(x) = 1, \qquad \rho_n \neq 0. \tag{9.11}$$

Then the corresponding moment functional $\mathfrak{M}$ is positive-definite on $[0, \infty]$ if and only if there are numbers δ_n such that

$$\delta_1 \geqq 0, \qquad \delta_n > 0, \qquad n \geqq 2,$$
$$f_n = \delta_{2n-1} + \delta_{2n}, \qquad \rho_{n+1} = \delta_{2n}\delta_{2n+1}, \qquad n \geqq 1. \tag{9.12}$$

Moreover, $\mathfrak{M} = \mathfrak{L}_0^*$ for some moment functional $\mathfrak{L}$ which is positive-definite on $[0, \infty]$ if and only if such δ_n exist with $\delta_1 > 0$.

Proof If $\delta_1 = 0$, set $\delta_n = \gamma_n$ $(n \geqq 2)$, $\rho_1 = \gamma_1$ and obtain $R_n(x) = P_n(x)$, $\mathfrak{M} = \mathfrak{L}$ as in Theorem 9.1.
   If $\delta_1 > 0$, set $\delta_n = \gamma_{n+1}$ $(n \geqq 1)$, $\rho_1 = \gamma_1\gamma_2$ so that $R_n(x) = Q_n(x)$, $\mathfrak{M} = \mathfrak{L}_0^*$ in Theorem 9.1. ∎
   Calling upon the Tchebichef polynomials for a final illustrative role, we take $S_n(x) = 2^{1-n}T_n(x)$ so that

$$\gamma_1 = \pi, \qquad \gamma_2 = \frac{1}{2}, \qquad \gamma_n = \frac{1}{4}, \qquad n \geqq 2.$$

We then obtain immediately from (9.9) and (9.10),

$$\lambda_1 = \pi, \qquad \nu_1 = \frac{\pi}{2}, \qquad c_1 = \frac{1}{2}, \qquad d_1 = \frac{3}{4},$$
$$c_n = d_n = \frac{1}{2}, \qquad \lambda_2 = \frac{1}{8}, \qquad \lambda_{n+1} = \nu_n = \frac{1}{16}, \qquad n \geqq 2, \tag{9.13}$$

as the coefficients in the recurrence formulas for $\{2^{1-2n}T_{2n}(x^{1/2})\}$ and $\{2^{-2n}x^{-1/2}T_{2n+1}(x^{1/2})\}$.
   In view of (8.6), (9.13) provides us with the recurrence formula for $\{V_n(x)\}$.
   The formulas in Theorem 9.1 can be used frequently (though less routinely) to obtain the recurrence formulas satisfied by the kernel and symmetric orthogonal polynomials corresponding to a given OPS whose

recurrence is known. For example, in the case of the monic Charlier polynomials, $\{C_n^{(a)}(x)\}$, we have (see Ex. 4.10)

$$c_n = n + a - 1, \qquad \lambda_{n+1} = an, \qquad n \geq 1, \qquad (\lambda_1 = e^a).$$

It is then easily deduced that (9.9) is satisfied by

$$\gamma_{2n} = a, \qquad \gamma_{2n+1} = n, \qquad n \geq 1.$$

In particular, for the kernel polynomials $Q_n(x) = C_n^*(0; x)$ $[C_n(x) = C_n^{(a)}(x)]$ the recurrence is

$$Q_{n+1}(x) = (x - n - a - 1)Q_n(x) - anQ_{n-1}(x), \qquad n \geq 0.$$

Comparison of this with the recurrence satisfied by $C_n^{(a)}(x)$ shows that $Q_n(x) = C_n^{(a)}(x - 1)$, a fact that could have been deduced by examining the "mass distribution" that defines the corresponding moment functional.

We conclude this introduction to the basic formal properties of orthogonal polynomials by recasting the corollary to Theorem 9.1 in a form which will be of great utility in our later analysis of the fundamental recurrence formula.

THEOREM 9.2 The true interval of orthogonality of $\mathfrak{M}$ is a subset of $[0, \infty]$ if and only if $f_n > 0$ for $n \geq 1$, and there are numbers g_n such that
(i)

$$0 \leq g_0 < 1, \qquad 0 < g_n < 1, \qquad n \geq 1;$$

(ii)

$$\frac{\rho_{n+1}}{f_n f_{n+1}} = (1 - g_{n-1})g_n, \qquad n = 1, 2, 3, \ldots.$$

Moreover, $\mathfrak{M} = \mathfrak{L}_0^*$ where $\mathfrak{L}$ is positive-definite on $[0, \infty]$ if and only if such g_n exist with $g_0 > 0$.

Proof If (9.12) is satisfied, then $f_n > \delta_{2n-1}$, hence

$$\delta_{2n-1} = g_{n-1}f_n$$

where $0 \leq g_0 < 1$ and $0 < g_n < 1$ for $n \geq 1$.

Then $\delta_{2n} = f_n - \delta_{2n-1} = (1 - g_{n-1})f_n$ so that we obtain (ii).

Conversely, if (i) and (ii) are satisfied, then (9.12) follows immediately upon putting $\delta_{2n-1} = g_{n-1}f_n$ and $\delta_{2n} = (1 - g_{n-1})f_n$.

Finally, we obviously have $\delta_1 > 0$ if and only if $g_0 > 0$. ∎

Exercises

9.1 Find explicitly the orthogonality relation for $\{F_n^{(a)}(x)\}$ satisfying

$$F_n^{(a)}(x) = xF_{n-1}^{(a)}(x) - \gamma_n F_{n-2}^{(a)}(x)$$

$$F_{-1}^{(a)}(x) = 0, \qquad F_0^{(a)}(x) = 1; \qquad \gamma_{2n} = a, \qquad \gamma_{2n+1} = n, \qquad n \geq 1.$$

9.2 Use the formulas of this section together with the recurrence formula for the Laguerre polynomials (Ex. 8.5) to obtain the recurrence formula for the generalized Hermite polynomials. Verify your result by deriving the same result using Theorem 4.2.

9.3 Obtain the recurrence formulas satisfied by $\{V_n(x)\}$ and $\{W_n(x)\}$ defined in (7.6) and Ex. 7.6, respectively.

9.4 Using the recurrence formula for the Legendre polynomials (Ex. 1.6), obtain the recurrence formulas for the monic orthogonal polynomials corresponding to the weight functions $x^{-1/2}$ and $x^{1/2}$, respectively, on $(0,1)$. From your result, derive the recurrence formulas for the special Jacobi polynomials, $\{P_n^{(0,-1/2)}(x)\}$ and $\{P_n^{(0,1/2)}(x)\}$.

9.5 Using the notation of Theorem 9.1, let, as usual, $p_n(x)$ denote the orthonormal polynomial corresponding to $P_n(x)$. Show that

$$[p_n(0)]^2 = \frac{\gamma_2 \gamma_4 \cdots \gamma_{2n}}{\gamma_1 \gamma_3 \cdots \gamma_{2n+1}}, \qquad n \geq 0.$$

9.6 Relative to Theorem 9.1, show that $P_n(0) \neq 0$ and obtain the formulas
(a)

$$\gamma_{2n} = -\frac{P_n(0)}{P_{n-1}(0)}, \qquad \gamma_{2n+1} = -\frac{\lambda_{n+1} P_{n-1}(0)}{P_n(0)}, \qquad n \geq 1;$$

(b)

$$P_n(x) = P_n^*(0; x) + \gamma_{2n+1} P_{n-1}^*(0; x), \qquad n \geq 0.$$

9.7 Assume that the conditions, (9.9) and (9.10) are satisfied with $\gamma_n > 0$. Let $\{x_{nk}\}_{k=1}^n$ and $\{y_{nk}\}_{k=1}^n$ denote the zeros of $P_n(x)$ and $Q_n(x)$, respectively. Prove that

$$\begin{aligned} \gamma_{2n} &= \sum_{k=1}^{n-1} (x_{n,k+1} - y_{n-1,k}) + x_{n1} \\ \gamma_{2n+1} &= \sum_{k=1}^{n} (y_{nk} - x_{nk}) \end{aligned} \qquad n \geq 1.$$

9.8 Let $\{R_n(x)\}$ satisfy (9.11) and assume that the conditions of Theorem 9.2 are satisfied. Suppose further that $g_0 > 0$. Let

$$\gamma_1 = \frac{\rho_1}{g_0 f_1}, \qquad \gamma_{2n} = g_{n-1} f_n, \qquad \gamma_{2n+1} = (1 - g_{n-1}) f_n, \qquad n \geq 1.$$

(a) Prove that $R_n(x) = P_n^*(0; x)$ where $\{P_n(x)\}$ satisfies (9.1) with c_n and λ_n defined by (9.9) with the above values of γ_n.

(b) Show that

$$\lambda_1[p_n(0)]^2 = \frac{g_0 g_1 \cdots g_{n-1}}{(1 - g_0)(1 - g_1) \cdots (1 - g_{n-1})}, \qquad n \geq 1.$$

9.9 In the study of certain Markov processes (birth-and-death processes), there arise polynomial sequences defined by recurrences of the form,

$$-xQ_n(x) = b_n Q_{n-1}(x) - (a_n + b_n)Q_n(x) + a_n Q_{n+1}(x), \qquad n \geq 0,$$

$$Q_{-1}(x) = 0, \qquad Q_0(x) = 1;$$

$$a_n > 0, \qquad b_{n+1} > 0 \quad \text{for } n \geq 0 \text{ and } b_0 \geq 0.$$

(a) Show that $\{Q_n(x)\}$ is an OPS with respect to a moment functional which is positive-definite on $[0, \infty]$.

(b) Prove that if $b_0 > 0$, then $\{Q_n(x)\}$ is a sequence of (non-monic) kernel polynomials with K-parameter 0 corresponding to the OPS $\{P_n(x)\}$ defined by

$$-xP_n(x) = a_{n-1}P_{n-1}(x) - (a_{n-1} + b_n)P_n(x) + b_n P_{n+1}(x), \qquad n \geq 0,$$

$$P_0(x) = b_0, \qquad a_{-1} = P_{-1}(x) = 0.$$

(Karlin and McGregor [1])

9.10 Let $\{P_n(x)\}$ be a monic OPS and suppose that $P_n^*(0; x) = P_n(x - c)$ for some $c \neq 0$. Prove that $P_n(x) = c^n C_n^{(a)}(x/c)$, where $C_n^{(a)}(x)$ denotes the monic Charlier polynomial (cf. Ex. 4.10).

9.11 Prove that if $\{P_n(x)\}$ is a monic OPS and $P_n^*(0; x) = q^{-n}P_n(qx)$, $q \neq 0, 1$, then

$$P_{n+1}(x) = \{x + [q - (1 + q)q^{-n}]q^{-n}b\} P_n(x) - (1 - q^n)q^{3-4n}b^2 P_{n-1}(x),$$

$$n \geq 0,$$

$$P_{-1}(x) = 0, \qquad P_0(x) = 1, \qquad b \neq 0.$$

$\{P_n(x)\}$ corresponds to a positive-definite moment functional if and only if $0 < q < 1$, b is real. The polynomials, $P_n(bq^{3/2}x)$, are constant multiples of the *Stieltjes-Wigert polynomials* (see VI-§2).

CHAPTER II

The Representation Theorem and Distribution Functions

1 Introduction

Let ψ be a function of bounded variation on $(-\infty, \infty)$ (that is, ψ is of bounded variation on every bounded interval $[a, b]$ and the corresponding total variations, $V_a^b(\psi)$, form a bounded set). If we set

$$\mathcal{L}[x^n] = \lim_{\substack{a \to -\infty \\ b \to +\infty}} \int_a^b x^n \, d\psi(x) = \int_{-\infty}^\infty x^n \, d\psi(x), \qquad n = 0, 1, 2, \ldots, \quad (1.1)$$

and if the improper integrals converge, then $\mathcal{L}$ is a moment functional.

In general, $\mathcal{L}$ will not be quasi-definite. However if ψ is non-decreasing, then $\mathcal{L}$ will be positive-definite if the set

$$\mathfrak{S}(\psi) = \{x \,|\, \psi(x + \delta) - \psi(x - \delta) > 0 \quad \text{for all } \delta > 0\} \quad (1.2)$$

is infinite. Moreover, $\mathfrak{S}(\psi)$ will be a supporting set for $\mathcal{L}$.

We now wish to show that the converse of this observation is true and thus characterize positive-definite moment functionals in terms of Stieltjes integrals with non-decreasing integrators.

DEFINITION 1.1 A bounded non-decreasing function ψ whose *moments*

$$\mu_n = \int_{-\infty}^\infty x^n \, d\psi(x), \qquad n = 0, 1, 2, \ldots,$$

are all finite is called a *distribution function*.

The set $\mathfrak{S}(\psi)$ defined by (1.2) is called the *spectrum* of ψ. A point in $\mathfrak{S}(\psi)$ is called a *spectral point* of ψ.

51

DEFINITION 1.2   Two distribution functions ψ_1 and ψ_2 are said to be *substantially equal* if and only if there is a constant C such that $\psi_1(x) = \psi_2(x) + C$ at all common points of continuity.

It is clear that substantially equal distribution functions have the same spectrum. Moreover, since changing the value of ψ at a point of discontinuity does not affect the value of the corresponding Stieltjes integral for continuous integrands, substantially equal distribution functions determine the same moment functional.

Finally, observe that $\mathfrak{S}(\psi)$ is a closed set of real numbers which is, in the terminology of measure theory, the support of the measure induced by ψ. In the language of the theory of distributions, it is the support of the distribution $d\psi$. Without assuming the latter theory, we will sometimes speak of the "distribution $x\,d\psi(x)$," as a convenient way of avoiding such involved statements as, "the distribution function, ϕ, defined by

$$\phi(x) = \int_{-\infty}^{x} t\,d\psi(t),"$$

when referring to kernel polynomials.

Exercises

1.1   Prove that the spectrum of a distribution function is a closed set.

1.2   Let $\mathcal{L}$ be defined by (1.1) where $\mathfrak{S}(\psi)$ is an infinite set. Prove that $\mathcal{L}$ is positive-definite on $\mathfrak{S}(\psi)$.

1.3   Prove that substantially equal distribution functions have the same points of continuity, the same jumps at points of discontinuity, and hence the same spectra.

2   Some Preliminary Theorems

We first establish some general convergence theorems which will be needed in the proof of the representation theorem. The first of these is frequently proven in courses in "modern" undergraduate analysis but we will include a proof for the sake of completeness.

THEOREM 2.1   Let $\{f_n\}$ be a sequence of real functions defined on a countable set E. If for each $x \in E$, $\{f_n(x)\}$ is bounded, then $\{f_n\}$ contains a subsequence that converges everywhere on E.

Proof   Let $E = \{x_1, x_2, x_3, \dots\}$ and   write $f_n^{(0)} = f_n$. Now   since

$\{f_n^{(0)}(x_1)\}$ is a bounded sequence of real numbers, it contains a convergent subsequence. That is, there is a sequence $\{f_n^{(1)}\}$ which is a subsequence of $\{f_n^{(0)}\}$ and which converges for $x = x_1$.

Now $\{f_n^{(1)}(x_2)\}$ is a bounded sequence so we can conclude as before that there is a subsequence $\{f_n^{(2)}\}$ of $\{f_n^{(1)}\}$ which converges for $x = x_2$. Continuing in this way, we obtain sequences $\{f_n^{(0)}\}, \{f_n^{(1)}\}, \ldots, \{f_n^{(i)}\}, \ldots$ such that

(a) $\{f_n^{(k)}\}$ is a subsequence of $\{f_n^{(k-1)}\}$ $(k = 1, 2, 3, \ldots)$;

(b) $\{f_n^{(k)}(x)\}$ converges for $x \in E_k = \{x_1, \ldots, x_k\}$.

It follows from (a) (with a little care to be certain that the relative order of terms is preserved) that the diagonal sequence, $\{f_n^{(n)}\}$, is a subsequence of $\{f_n\}$. Since, except possibly for the first $k - 1$ terms, $\{f_n^{(n)}\}$ is also a subsequence of $\{f_n^{(k)}\}$, it follows from (b) that $\{f_n^{(n)}(x)\}$ converges for $x \in E = \cup_{k=1}^{\infty} E_k$. ∎

We next prove a theorem which, when stated in terms of functions of bounded variation, is usually known as *Helly's Selection Principle* (or *Theorem of Choice*). For the sake of brevity, we state it for non-decreasing functions although there is no difficulty involved in the proof of the more general case.

THEOREM 2.2 Let $\{\phi_n\}$ be a uniformly bounded sequence of non-decreasing functions defined on $(-\infty, \infty)$. Then $\{\phi_n\}$ has a subsequence which converges on $(-\infty, \infty)$ to a bounded, non-decreasing function.

Proof Let Q denote the rational numbers. According to Theorem 2.1 there is a subsequence $\{\phi_{n_k}\}$ which converges everywhere on Q. We then define a function ϕ^* on Q by

$$\phi^*(r) = \lim_{k \to \infty} \phi_{n_k}(r), \qquad r \in Q.$$

It follows from the hypothesis on $\{\phi_n\}$ that ϕ^* is bounded and non-decreasing on Q. We now extend the domain of ϕ^* to $\mathcal{R} = (-\infty, \infty)$ by defining

$$\phi^*(x) = \sup\{\phi^*(r) | r \in Q \quad \text{and} \quad r < x\}, \qquad x \in \mathcal{R} \backslash Q.$$

ϕ^* is clearly bounded and non-decreasing on $\mathcal{R}$. We next show that $\{\phi_{n_k}(x)\}$ converges to $\phi^*(x)$ at all points x of continuity of ϕ^*.

To this end, suppose ϕ^* is continuous at $x \notin Q$. Since Q is dense in $\mathcal{R}$, then given $\epsilon > 0$, there is an $x_2 > x$, $x_2 \in Q$, such that

$$\phi^*(x_2) \leqq \phi^*(x) + \epsilon.$$

For any $x_1 \in Q$, $x_1 < x$, we also have

$$\phi_{n_k}(x_1) \leqq \phi_{n_k}(x) \leqq \phi_{n_k}(x_2).$$

Thus

$$\phi^*(x_1) \leqq \liminf_{k \to \infty} \phi_{n_k}(x) \leqq \limsup_{k \to \infty} \phi_{n_k}(x) \leqq \phi^*(x_2).$$

Therefore

$$\phi^*(x) \leqq \liminf_{k \to \infty} \phi_{n_k}(x) \leqq \limsup_{k \to \infty} \phi_{n_k}(x) \leqq \phi^*(x) + \epsilon,$$

whence it follows that $\{\phi_{n_k}\}$ converges to ϕ^* at all points of continuity of ϕ^*.

But ϕ^* is non-decreasing so its points of discontinuity form a countable set D. Applying Theorem 2.1 to $\{\phi_{n_k}\}$ and D, we conclude that there is a subsequence of $\{\phi_{n_k}\}$ which converges on D, hence on $\mathcal{R}$, to a limit function ϕ. (ϕ is of course identical with ϕ^* on $\mathcal{R} \backslash D$.) The hypotheses on $\{\phi_n\}$ guarantee that ϕ is bounded and non-decreasing. ∎

As our final preliminary result, we prove *Helly's second theorem*. As before, we consider non-decreasing functions although there is no difficulty involved in treating the more general case involving functions of bounded variation.

THEOREM 2.3 Let $\{\phi_n\}$ be a uniformly bounded sequence of non-decreasing functions defined on a compact interval $[a, b]$, and let it converge on $[a, b]$ to a limit function ϕ. Then for every real function f continuous on $[a, b]$,

$$\lim_{n \to \infty} \int_a^b f \, d\phi_n = \int_a^b f \, d\phi.$$

Proof Since $\{\phi_n\}$ is uniformly bounded, there exists an $M > 0$ such that

$$0 \leqq \phi_n(b) - \phi_n(a) \leqq M, \qquad n = 1, 2, 3, \ldots,$$

hence also

$$0 \leqq \phi(b) - \phi(a) \leqq M.$$

Let $\epsilon > 0$ be given. If f is real and continuous on $[a, b]$, then f is uniformly continuous on $[a, b]$ so there is a partition $P_\epsilon = \{x_0, x_1, \ldots, x_m\}$ of $[a, b]$ such that

$$|f(x') - f(x'')| < \epsilon \qquad \text{for } x', x'' \in [x_{i-1}, x_i], \qquad 1 \leqq i \leqq m.$$

Choose $\xi_i \in [x_{i-1}, x_i]$ and write

$$\Delta_i \phi = \phi(x_i) - \phi(x_{i-1}), \qquad \Delta_i \phi_n = \phi_n(x_i) - \phi_n(x_{i-1}).$$

By the mean value theorem for Stieltjes integrals,

$$\int_{x_{i-1}}^{x_i} f d\phi - f(\xi_i)\Delta_i \phi = [f(\xi_i') - f(\xi_i)]\Delta_i \phi$$

for some $\xi_i' \in [x_{i-1}, x_i]$. Summing over i, we obtain

$$\left| \int_a^b f d\phi - \sum_{i=1}^m f(\xi_i)\Delta_i \phi \right| \leq \sum_{i=1}^m |f(\xi_i') - f(\xi_i)|\Delta_i \phi$$

$$< \epsilon \sum_{i=1}^m \Delta_i \phi \leq \epsilon M.$$

In the same way we find

$$\left| \int_a^b f d\phi_n - \sum_{i=1}^m f(\xi_i)\Delta_i \phi_n \right| < \epsilon M.$$

Therefore

$$\left| \int_a^b f d\phi - \int_a^b f d\phi_n \right| \leq \left| \int_a^b f d\phi - \sum_{i=1}^m f(\xi_i)\Delta_i \phi \right| + \left| \sum_{i=1}^m f(\xi_i)(\Delta_i \phi - \Delta_i \phi_n) \right|$$

$$+ \left| \int_a^b f d\phi_n - \sum_{i=1}^m f(\xi_i)\Delta_i \phi_n \right|$$

$$< 2M\epsilon + \sum_{i=1}^m |f(\xi_i)||\Delta_i(\phi - \phi_n)|.$$

Keeping P_ϵ fixed, we have $\lim_{n\to\infty} \Delta_i(\phi - \phi_n) = 0$, hence

$$\limsup_{n\to\infty} \left| \int_a^b f d\phi - \int_a^b f d\phi_n \right| \leq 2M\epsilon$$

and the desired conclusion follows.   ∎

Exercises

2.1 Let $\{c_{nk}\}_{n=1}^\infty$ be a bounded sequence for each fixed positive integer k. Show that there are integers $1 \leq n_1 < n_2 < n_3 < \cdots$ such that $\{c_{n_i k}\}_{i=1}^\infty$ converges for every k.

2.2 State and prove the two Helly theorems for the case where the ϕ_n are all functions of bounded variation.

3 The Representation Theorem

Let $\mathcal{L}$ be a positive-definite moment functional with moment sequence $\{\mu_n\}_{n=0}^{\infty}$. According to the Gauss quadrature formula (I-Theorem 6.1), for each positive integer n, there are positive numbers $A_{n1}, \ldots, A_{nn}$ such that

$$\mathcal{L}[x^k] = \mu_k = \sum_{i=1}^{n} A_{ni} x_{ni}^k, \qquad k = 0, 1, \ldots, 2n - 1, \qquad (3.1)$$

where $x_{n1} < x_{n2} < \cdots < x_{nn}$ are the zeros of $P_n(x)$, the nth degree monic orthogonal polynomial corresponding to $\mathcal{L}$.

Let ψ_n be defined by

$$\psi_n(x) = \begin{cases} 0 \text{ if } x < x_{n1} \\ A_{n1} + \cdots + A_{np} & \text{if } x_{np} \leqq x < x_{n,p+1} \quad (1 \leqq p < n) \\ \mu_0 \text{ if } x \geqq x_{nn}. \end{cases} \qquad (3.2)$$

ψ_n is a bounded, right continuous, non-decreasing step function whose spectrum is the finite set $\{x_{n1}, \ldots, x_{nn}\}$, and whose jump at x_{ni} is $A_{ni} > 0$. Thus

$$\int_{-\infty}^{\infty} x^k \, d\psi_n(x) = \sum_{i=1}^{n} A_{ni} x_{ni}^k = \mu_k, \qquad k = 0, 1, \ldots, 2n - 1. \quad (3.3)$$

According to Theorem 2.2, $\{\psi_n\}$ contains a subsequence which converges on $(-\infty, \infty)$ to a bounded, non-decreasing function ψ. If $[\xi_1, \eta_1]$, the true interval of orthogonality of $\mathcal{L}$, was bounded, then Theorem 2.3 could be invoked to conclude from (3.3) that

$$\int_{-\infty}^{\infty} x^k \, d\psi(x) = \mu_k = \mathcal{L}[x^k], \qquad k = 0, 1, 2, \ldots \qquad (3.4)$$

(Note that $\psi(x) = 0$ for $x \leqq \xi_1$ and $\psi(x) = \mu_0$ for $x \geqq \eta_1$ so (3.4) can be written with the integration interval reduced to $[\xi_1, \eta_1]$.)

In the event $[\xi_1, \eta_1]$ is unbounded, one might look for an extension of Helly's second theorem to unbounded intervals. Unfortunately, the "obvious" or "natural" extension is not valid. For example, if

$$\psi_n(x) = \begin{cases} 0 \text{ for } x < n \\ 1 \text{ for } x \geqq n, \end{cases} \qquad n = 1, 2, 3, \ldots \qquad (3.5)$$

then $\{\psi_n\}$ is a uniformly bounded sequence of non-decreasing functions which converges everywhere to $\psi(x) = 0$. However,

$$\int_0^{\infty} x^k \, d\psi_n(x) = n^k \quad \text{while} \quad \int_0^{\infty} x^k \, d\psi(x) = 0.$$

We therefore give a direct proof that (3.4) is valid.

THEOREM **3.1**   Let $\mathfrak{L}$ be a positive-definite moment functional and let ψ_n be defined by (3.2). Then there is a subsequence of $\{\psi_n\}$ that converges on $(-\infty, \infty)$ to a distribution function ψ which has an infinite spectrum and for which (3.4) is valid.

Proof   As already noted, there is a subsequence $\{\psi_{n_i}\}$ which converges on $(-\infty, \infty)$ to a distribution function ψ. Writing $\phi_i = \psi_{n_i}$, we have according to (3.3)

$$\int_{-\infty}^{\infty} x^k \, d\phi_i(x) = \mu_k \qquad \text{for } n_i \geqq \frac{k+1}{2}.$$

From Theorem 2.3 we conclude that for every compact interval $[\alpha, \beta]$,

$$\lim_{i \to \infty} \int_{\alpha}^{\beta} x^k \, d\phi_i(x) = \int_{\alpha}^{\beta} x^k \, d\psi(x). \tag{3.6}$$

Choosing $\alpha < 0 < \beta$ and $n_i > k + 1$, we can write

$$|\mu_k - \int_{\alpha}^{\beta} x^k \, d\psi(x)| = |\int_{-\infty}^{\infty} x^k \, d\phi_i(x) - \int_{\alpha}^{\beta} x^k \, d\psi(x)|$$

$$\leqq |\int_{-\infty}^{\alpha} x^k \, d\phi_i(x)| + |\int_{\beta}^{\infty} x^k \, d\phi_i(x)| + |\int_{\alpha}^{\beta} x^k \, d\phi_i(x) - \int_{\alpha}^{\beta} x^k \, d\psi(x)|.$$

But

$$|\int_{\beta}^{\infty} x^k \, d\phi_i(x)| = \left| \int_{\beta}^{\infty} \frac{x^{2k+2}}{x^{k+2}} \, d\phi_i(x) \right| \leqq \beta^{-(k+2)} |\int_{\beta}^{\infty} x^{2k+2} \, d\phi_i(x)|$$

$$\leqq \beta^{-(k+2)} \mu_{2k+2}.$$

Similarly

$$|\int_{-\infty}^{\alpha} x^k \, d\phi_i(x)| \leqq |\alpha|^{-(k+2)} \mu_{2k+2},$$

so that

$$|\mu_k - \int_{\alpha}^{\beta} x^k \, d\psi(x)| \leqq |\int_{\alpha}^{\beta} x^k \, d\phi_i(x) - \int_{\alpha}^{\beta} x^k \, d\psi(x)|$$

$$+ (|\alpha|^{-k-2} + \beta^{-k-2}) \mu_{2k+2}.$$

Hence letting $i \to \infty$, we have by (3.6)

$$\left| \mu_k - \int_\alpha^\beta x^k \, d\psi(x) \right| \leqq (|\alpha|^{-k-2} + \beta^{-k-2})\mu_{2k+2}.$$

Now if we let $\alpha \to -\infty$ and $\beta \to \infty$, we obtain (3.4).

Finally it is easy to show that ψ has an infinite spectrum. For if the spectrum of ψ consisted of exactly N points, we could construct a real polynomial $\pi(x)$ which vanished at these N points. We would then have

$$\mathcal{L}[\pi^2(x)] = \int_{-\infty}^{\infty} \pi^2(x) \, d\psi(x) = 0,$$

contradicting the positive-definiteness of $\mathcal{L}$. ∎

We have thus shown that every positive-definite moment functional can be represented as a Stieltjes integral with a non-decreasing integrator ψ whose spectrum is an infinite set. We will say that "ψ provides a representation for $\mathcal{L}$" or simply that "ψ is a *representative* of $\mathcal{L}$." Various other obvious grammatical variants of this will also be used.

It has already been noted that a representative ψ for $\mathcal{L}$ cannot be unique since ψ can be replaced by another distribution function which is substantially equal to ψ. It must not be inferred at this point however that $\mathcal{L}$ determines a unique equivalence class of substantially equal distribution functions. For it can happen that the same moment functional $\mathcal{L}$ can be represented by (infinitely many) substantially unequal distribution functions. Of course these substantially unequal distribution functions would define different extensions of $\mathcal{L}$ to linear functionals defined over a wider class of integrable functions than the polynomials.

DEFINITION 3.1 A positive-definite moment functional $\mathcal{L}$ is called *determinate* if any two representatives of $\mathcal{L}$ are substantially equal. (That is, $\mathcal{L}$ has a *substantially unique* representative.) Otherwise, $\mathcal{L}$ is called *indeterminate*.

The question of when $\mathcal{L}$ is determinate is a difficult one which we will discuss briefly in §5. Although we will not obtain criteria for determinacy in the general case, we will show that if $\mathcal{L}$ has a bounded supporting set, then $\mathcal{L}$ is determinate. For now we will content ourselves with the following partial result.

THEOREM 3.2 Every positive-definite moment functional $\mathcal{L}$ has a representative whose spectrum is a subset of $[\xi_1, \eta_1]$. Further, $[\xi_1, \eta_1]$ is a subset of every closed interval that contains the spectrum of some representative of $\mathcal{L}$.

Proof Let ψ_n be defined by (3.2) and let ψ denote the corresponding representative for $\mathfrak{L}$ which is a subsequential limit of $\{\psi_n\}$. It is then clear that if $\xi_1 > -\infty$, $\psi(x) = 0$ for $x \leqq \xi_1$ while if $\eta_1 < \infty$, then $\psi(x) = \mu_0$ for $x \geqq \eta_1$. The first assertion follows.

Now if ϕ is a representative for $\mathfrak{L}$ and $\mathfrak{S}(\phi) \subset [a,b]$, then $[a,b]$ is a supporting set for $\mathfrak{L}$. By I-Theorem 5.2, $[\xi_1, \eta_1] \subset [a,b]$. ∎

Exercises

3.1 Give an example of a sequence of non-decreasing functions, ϕ_n, each having an infinite spectrum and finite moments and such that $\{\phi_n\}$ converges to ϕ but $\int_0^\infty x^k d\phi_n(x)$ does not converge to $\int_0^\infty x^k d\phi(x)$.

3.2 (The following assumes a knowledge of functional analysis.) Let $X = [\xi_1, \eta_1]$ be a compact interval and let $C(X)$ denote the space of all continuous functions on X.

  (i) Show that the Hahn-Banach theorem can be applied to obtain an extension of $\mathfrak{L}$ to a linear functional $\bar{\mathfrak{L}}$ defined on $C(X)$. (cf. I-Ex. 5.6)

  (ii) Show that $\bar{\mathfrak{L}}$ is a positive functional on $C(X)$ and use the Riesz representation theorem to obtain the representation theorem for $\mathfrak{L}$.

4 Spectral Points and Zeros of Orthogonal Polynomials

We maintain the hypothesis that $\mathfrak{L}$ is positive-definite and continue to denote by $\{P_n(x)\}$ the corresponding monic OPS and by $\{x_{ni}\}_{i=1}^n$ the zeros of $P_n(x)$ arranged by increasing size.

According to Theorem 3.1, the sequence of step functions $\{\psi_n\}$ contains a subsequence converging to a representative ψ of $\mathfrak{L}$. Since the jumps of ψ_n occur at the zeros of $P_n(x)$, this indicates there is an intimate relationship between the spectrum of ψ and the distribution of zeros of the $P_n(x)$. We shall obtain a few of these relations here including some that apply generally to all representatives in the event $\mathfrak{L}$ is indeterminate. More detailed results are available under special hypotheses on a given distribution function while more complete general information regarding the nature of the spectra requires a thorough study of the "indeterminate Hamburger moment problem." For these, see for example Szegö [5] and Shohat and Tamarkin [1], respectively.

THEOREM 4.1 Let ϕ be any representative of $\mathfrak{L}$. Then the spectrum of ϕ contains at least one point in each open interval, $(x_{ni}, x_{n,i+1})$, $1 \leqq i \leqq n$; $n = 2, 3, \ldots$.

Proof Assume that $\mathfrak{S}(\phi) \cap (x_{nk}, x_{n,k+1}) = \varnothing$ for some $n \geqq 2$, $1 \leqq k \leqq n$. Put

$$\pi(x) = \frac{P_n(x)}{(x - x_{nk})(x - x_{n,k+1})}$$

so that $\pi(x)$ is a polynomial of degree $n - 2 \geqq 0$, not identically zero, such that $\pi(x) P_n(x) \geqq 0$ for $x \notin (x_{nk}, x_{n,k+1})$.

In particular, $\pi(x) P_n(x) \geqq 0$ for $x \in \mathfrak{S}(\phi)$. Since $\mathfrak{S}(\phi)$ is a supporting set for $\mathfrak{L}$, $\mathfrak{L}[\pi(x) P_n(x)] > 0$, and this contradicts the orthogonality property of $P_n(x)$. ∎

An alternate statement of Theorem 4.1 is of course that any interval of constancy of ϕ contains at most one zero of any $P_n(x)$. On the other hand, we will show that any open interval that is "eventually free of zeros" is an interval of constancy for the representative ψ given by Theorem 3.1. It will therefore be convenient to give this representative a name.

DEFINITION 4.1 A distribution function ψ is called a *natural representative* of $\mathfrak{L}$ if ψ is a subsequential limit of $\{\psi_n\}$ (defined by (3.2)).

THEOREM 4.2 Let G be any open set such that for all n sufficiently large, $P_n(x)$ has no zero in G. Then a natural representative ψ of $\mathfrak{L}$ has no spectral point in G. (That is, ψ is constant on each connected subset of G.)

Proof If I is an open interval, $I \subset G$, then there exists N such that $P_n(x)$ has no zero in I for $n \geqq N$. Thus if ψ_n is defined by (3.2), then

$$\psi_n(x') - \psi_n(x'') = 0 \qquad \text{for } x', x'' \in I, \quad n \geqq N.$$

Hence $\psi(x') - \psi(x'') = 0$. ∎

As an immediate corollary to Theorem 4.2, we have the following important result.

THEOREM 4.3 Let ψ be a natural representative of $\mathfrak{L}$, and let $s \in \mathfrak{S}(\psi)$. Then every neighborhood of s contains a zero of $P_n(x)$ for infinitely many values of n.

Proof This is a restatement of Theorem 4.2 in contrapositive form. ∎

Note that Theorem 4.3 says that if s is a spectral point of a natural representative of $\mathfrak{L}$, then either s is a limit point of the set

$$X = \{x_{ni} | 1 \leqq i \leqq n; \; n = 1, 2, 3, \ldots\} \tag{4.1}$$

or s is itself a zero of $P_n(x)$ for infinitely many values of n.

In other words, if we denote the derived set of X by X' and write

$$Z = \{x | P_n(x) = 0 \text{ for infinitely many } n\} \tag{4.2}$$

then if ψ is a natural representative of $\mathfrak{L}$,

$$\mathfrak{S}(\psi) \subset X' \cup Z. \tag{4.3}$$

Equality need not hold in (4.3) (see Ex. 4.2).

We next recall the limits

$$\xi_i = \lim_{n \to \infty} x_{ni}, \qquad \eta_j = \lim_{n \to \infty} x_{n,n-j+1}, \qquad i, j = 1, 2, 3, \ldots \tag{4.4}$$

We have (in the extended real number system), $\xi_{i-1} \leqq \xi_i < \eta_j \leqq \eta_{j-1}$, and hence define

$$\sigma = \begin{cases} -\infty & \text{if } \xi_i = -\infty & \text{for all } i \\ \lim_{i \to \infty} \xi_i & \text{if } \xi_p > -\infty & \text{for some } p \end{cases} \tag{4.5}$$

$$\tau = \begin{cases} +\infty & \text{if } \eta_j = +\infty & \text{for all } j \\ \lim_{j \to \infty} \eta_j & \text{if } \eta_q < +\infty & \text{for some } q. \end{cases} \tag{4.6}$$

Finally we write

$$\xi_0 = -\infty, \qquad \eta_0 = +\infty \; \_$$

so that

$$-\infty = \xi_0 \leqq \xi_1 \leqq \xi_2 \leqq \cdots \leqq \sigma \leqq \tau \leqq \cdots \leqq \eta_2 \leqq \eta_1 \leqq \eta_0 = +\infty. \tag{4.7}$$

THEOREM 4.4 Let ϕ be an arbitrary representative of $\mathfrak{L}$.

(i) If $\xi_k < \xi_{k+1}$ for some $k \geqq 0$, then

$$\mathfrak{S}(\phi) \cap (\xi_k, \xi_{k+1}] \neq \varnothing;$$

(ii) If $\xi_k = \xi_{k+1}$ for some $k \geqq 0$, then ξ_k is a limit point of $\mathfrak{S}(\phi)$;

(iii) σ is a limit point of $\mathfrak{S}(\phi)$.

Proof Suppose $\xi_k < \xi_{k+1}$. If $k = 0$, the conclusion of (i) follows from Theorem 3.2. If $k \geqq 1$, then in view of the definition of ξ_k, it follows that corresponding to a given $\epsilon > 0$, there is an N such that

$$(x_{nk}, x_{n,k+1}) \subset (\xi_k, \xi_{k+1} + \epsilon) \qquad \text{for } n \geqq N.$$

Theorem 4.1 now yields (i).

Next suppose $\xi_k = \xi_{k+1}$. If $k = 0$, then $\mathfrak{S}(\phi)$ is unbounded on the left so ξ_0 is a limit point of $\mathfrak{S}(\phi)$. If $k \geqq 1$, we again call on Theorem 4.1 and conclude that there is a sequence of distinct spectral points converging on ξ_k from the right.

Finally (iii) follows from (i) and (ii).   ■

THEOREM 4.5 Let ψ be a natural representative of $\mathfrak{L}$. If $\xi_1 > -\infty$, then $\xi_i \in \mathfrak{S}(\psi)$ for every $i > 0$ and $\mathfrak{S}(\psi)$ contains no points smaller than σ except, possibly, the ξ_i.

Proof If $\xi_i = \xi_{i+1} > -\infty$ for some $i > 0$, the asserted conclusion follows from Theorem 4.4 since $\mathfrak{S}(\psi)$ is a closed set of real numbers.

On the other hand, if $\xi_p < \xi_{p+1}$ for some $p \geqq 0$, then for every α, $\xi_p < \alpha < \xi_{p+1}$, the open interval (α, ξ_{p+1}) is "eventually free of zeros." Hence by Theorem 4.3, (α, ξ_{p+1}) contains no points of $\mathfrak{S}(\psi)$. Therefore we see that $(\xi_p, \xi_{p+1}) \cap \mathfrak{S}(\psi) = \varnothing$, and hence by Theorem 4.4 that $\xi_{p+1} \in \mathfrak{S}(\psi)$.

Finally, if $x < \sigma$, $x \neq \xi_i$ for every i, then $\xi_p < x < \xi_{p+1}$ for some $p \geqq 0$ so by Theorem 4.3 we can conclude that $x \notin \mathfrak{S}(\psi)$.   ■

THEOREM 4.6 If $\xi_p = \xi_{p+1}$ for some $p \geqq 0$, then $\xi_p = \sigma$.

Proof Suppose $\xi_1 = -\infty$ and assume $\sigma > -\infty$. Then there is a $p \geqq 1$ such that $-\infty = \xi_p < \xi_{p+1} \leqq \sigma$. It would then follow from Theorem 4.2 that a natural representative ψ has no spectral point in $(-\infty, \xi_{p+1})$. But this contradicts Theorem 3.2.

Next suppose $\xi_p = \xi_{p+1}$ for some $p \geqq 1$. If $\xi_1 = -\infty$, the previous argument shows $\sigma = \xi_p = -\infty$. If $\xi_1 > -\infty$, then Theorem 4.5 shows that the spectrum of a natural representative has no limit points smaller than σ. But according to Theorem 4.4, ξ_p is a limit point of the spectrum of every representative. Since σ is also a limit point, $\sigma = \xi_p$.   ■

In summary, we observe that we have only the following possibilities regarding the behavior of the ξ_i:

(a) $\xi_i = \sigma = -\infty$ ($i \geqq 1$);

(b) $-\infty < \xi_1 < \xi_2 < \cdots < \xi_p = \sigma$ for some $p \geqq 1$;

(c) $-\infty < \xi_1 < \xi_2 < \cdots < \xi_i < \xi_{i+1} < \cdots < \sigma$.

In case (c), if we also have $\sigma = +\infty$, then the spectrum of some natural representative ψ is precisely the set

$$\Xi = \{\xi_i | i = 1, 2, 3, \ldots\}.$$

Now it is obvious that analogous results are valid for the points η_j and τ. (These could be obtained immediately by considering $\mathcal{L}^-$ defined by $\mathcal{L}^-[x^n] = (-1)^n \mathcal{L}[x^n]$.) Therefore if we write

$$H = \{\eta_j | j = 1, 2, 3, \ldots\},$$

then it is easy to see that when σ and τ are *both finite*,

$$\mathfrak{S}(\psi) = \Xi \cup \mathfrak{S}_1 \cup H$$

where $\mathfrak{S}_1 \subset [\sigma, \tau]$. We remark that in this case, $\mathcal{L}$ will actually be determinate. In fact, it can be shown that $\mathcal{L}$ is determinate if either σ or τ is finite.

Exercises

4.1 Assume $-\infty < \xi_1 < \xi_2 < \cdots < \xi_i < \xi_{i+1} < \cdots$. Prove that

$$0 < \psi(\xi_i) - \psi(\xi_i - 0) \leqq \left\{ \sum_{k=0}^{\infty} p_k^2(\xi_i) \right\}^{-1}.$$

(Use I-Ex. 6.2)

4.2 If $\mathcal{S}$ is a symmetric positive-definite functional, then 0 is a zero of infinitely many of the corresponding orthogonal polynomials. Show that, however, there are such $\mathcal{S}$ for which 0 does not belong to the spectrum of a natural representative.

4.3 Prove that if $\mathcal{S}$ is a symmetric positive-definite moment functional, then the spectrum of a natural representative is symmetric about the origin. (This need not be true for the spectrum of every representative. See Ex. 6.2.)

4.4 Assume $0 < \xi_i < \xi_{i+1}$ $(i \geqq 1)$ and $\sum_{i=1}^{\infty} \xi_i^{-1} < \infty$. Prove that $\{P_n(z)/P_n(0)\}_{n=0}^{\infty}$ is uniformly bounded on every bounded subset of the complex plane.

4.5 Prove the following generalization of Theorem 4.1. Let F be a closed supporting set for $\mathcal{L}$ and let $q(x)$ be a real polynomial of degree at most $n + 1$ such that $\mathcal{L}[x^k q(x)] = 0$ for $0 \leqq k < n$. If z_1 and z_2 are real consecutive zeros of $q(x)$, $F \cap (z_1, z_2) \neq \varnothing$.

5 Determinacy of $\mathcal{L}$ In the Bounded Case

We will now show that if the true interval of orthogonality $[\xi_1, \eta_1]$ of the positive-definite moment functional $\mathcal{L}$ is bounded, then $\mathcal{L}$ is determinate. No attempt will be made to develop criteria for this question of determinacy in the general case. However our approach will provide the elementary tools which can form the basis of an investigation into this central question of the "problem of moments."

We begin with a concept introduced by M. Riesz in 1921 as the basis of his penetrating analysis of the moment problem.

DEFINITION 5.1   A polynomial $q(x)$, not identically zero, is called a *quasi-orthogonal polynomial of order* $n + 1$ if and only if it is of degree at most $n + 1$ and

$$\mathcal{L}[x^k q(x)] = 0 \qquad \text{for } k = 0, 1, \ldots, n - 1. \tag{5.1}$$

Note that according to this definition, the orthogonal polynomials $P_n(x)$ and $P_{n+1}(x)$ are both quasi-orthogonal polynomials of order $n + 1$.

THEOREM 5.1   (i) $q(x)$ is a quasi-orthogonal polynomial of order $n + 1$ if and only if there are constants A and B, not both zero, such that

$$q(x) = AP_{n+1}(x) + BP_n(x). \tag{5.2}$$

(ii) For each number z_0, there is a quasi-orthogonal polynomial of order $n + 1$, $q(x)$, such that $q(z_0) = 0$. This $q(x)$ is uniquely determined up to an arbitrary non-zero factor, and its degree is $n + 1$ if and only if $P_n(z_0) \neq 0$.

Proof   It is clear from (5.1) that $AP_{n+1}(x) + BP_n(x)$ is a quasi-orthogonal polynomial of order $n + 1$ provided $|A| + |B| \neq 0$. Conversely, if $q(x)$ is a quasi-orthogonal polynomial of order $n + 1$, we can write

$$q(x) = \sum_{k=0}^{n+1} c_k P_k(x)$$

where $c_k = \{\mathcal{L}[P_k^2(x)]\}^{-1} \mathcal{L}[q(x)P_k(x)] = 0$ for $0 \leq k < n - 1$.

Turning to (ii), we note that $P_n(z_0)$ and $P_{n+1}(z_0)$ cannot both vanish. Thus we can determine constants A and B, not both zero, such that $AP_{n+1}(z_0) + BP_n(z_0) = 0$. The proof can now be completed by equally elementary arguments. ∎

It is easily concluded from the preceding that real quasi-orthogonal polynomials have real, simple zeros for which a number of separation properties hold. We leave these for the reader (for example, see Ex. 5.2 and also I-Ex. 5.4) and state only the following partial result which will be used in the proof of Theorem 5.5.

THEOREM 5.2   The zeros of a real quasi-orthogonal polynomial are all real and simple. At most one of these zeros lies outside the open interval, (ξ_1, η_1).

Proof   If $q(x)$ is an orthogonal polynomial, there is nothing new to prove. Therefore assume $q(x) = AP_{n+1}(x) + BP_n(x)$ where A and B are real and different from zero.

With the usual meaning for x_{nk}, we have $q(x_{n+1,i}) = BP(x_{n+1,i})$ so $q(x_{n+1,i})$ alternates in sign as i varies from 1 to $n + 1$. That is, $q(x)$ has n real zeros separating the $n + 1$ zeros of $P_{n+1}(x)$. Since $q(x)$ is real, its remaining zero must be real and must lie outside $[x_{n+1,1}, x_{n+1,n+1}]$. ∎

Quasi-orthogonal polynomials can be used to obtain a simple generalization of the Gauss quadrature formula.

THEOREM 5.3   Let $q(x)$ be a *real* quasi-orthogonal polynomial of order and degree $n + 1$ and let $y_{n0} < y_{n1} < \cdots < y_{nn}$ denote its zeros. Then there are positive numbers B_{ni}, depending only on $q(x)$, such that

$$\mathcal{L}[\pi(x)] = \sum_{i=0}^{n} B_{ni}\pi(y_{ni}), \qquad (5.3)$$

for every polynomial of degree at most $2n$.

Except for a trivial modification relative to the degree of precision of (5.3), the proof is identical with that of I-Theorem 6.1 and is therefore omitted. It should be noted that if $q(x) = AP_{n+1}(x)$, then (5.3) is a Gauss quadrature formula and is then valid for polynomials of degree at most $2n + 1$.

As is the case with the Gauss quadrature formula, we can write

$$\mathcal{L}[x^k] = \int_{-\infty}^{\infty} x^k \, d\phi_n(x), \qquad k = 0, 1, \ldots, 2n \qquad (5.4)$$

where ϕ_n is a step function which has jumps of magnitude B_{ni} at y_{ni}. If we fix a real number x_0 which is not a zero of any $P_n(x)$ (at least for n sufficiently large), then we can construct a sequence of real quasi-orthogonal polynomials of order $n + 1$ ($n = 1, 2, 3, \ldots$) all of which vanish at x_0. Corresponding to this sequence, we obtain a sequence of distribution functions $\{\phi_n\}$ and each ϕ_n satisfies (5.4) and has x_0 in its spectrum.

Then as before we could use Helly's selection principle to extract a subsequence converging to a distribution function which could then be proved to be a representative for $\mathfrak{L}$. It is obvious that a study of such sequences would be of fundamental importance in the study of the question of determinacy of $\mathfrak{L}$.

We will not pursue this viewpoint here. Instead we now present an ad hoc proof of the determinacy of $\mathfrak{L}$ when the true interval of orthogonality is bounded. As a preliminary, we first note the following important extremal property of quasi-orthogonal polynomials.

THEOREM 5.4 Let x_0 be any real number which is not a zero of $P_n(x)$. Let $q(x)$ denote a real quasi-orthogonal polynomial of order and degree $n + 1$ which vanishes at x_0. If Λ_{n0} denotes the quadrature coefficient in (5.3) which corresponds to x_0, then

$$\Lambda_{n0} = \min \mathfrak{L}[|\pi(x)|^2]$$

where the minimum is computed as $\pi(x)$ ranges over all polynomials of degree at most n such that $\pi(x_0) = 1$.

Proof Referring to (5.3), suppose that $y_{np} = x_0$ and write $B_{np} = \Lambda_{n0}$. If $\pi(x)$ has degree not exceeding n and $\pi(x_0) = 1$, then

$$\mathfrak{L}[|\pi(x)|^2] \geqq B_{np}|\pi(x_0)|^2 = \Lambda_{n0}.$$

On the other hand, the polynomial

$$\rho(x) = \frac{q(x)}{(x - x_0)q'(x_0)} \tag{5.5}$$

is of degree n, vanishes at y_{ni} for $i \neq p$, and $\rho(x_0) = 1$. That is, $\mathfrak{L}[\rho^2(x)] = \Lambda_{n0}$. ∎

COROLLARY

$$\Lambda_{n0} = \left\{ \sum_{k=0}^{n} p_k^2(x_0) \right\}^{-1}$$

(where $p_k(x)$ again denotes the kth orthonormal polynomial).

Proof Compare Theorem 5.4 with I-Theorem 7.3 and I-(7.8). ∎

We next obtain a special case of the celebrated *Tchebichef inequalities* (see Ex. 5.6). We will write

$$\phi(-\infty) = \lim_{x \to -\infty} \phi(x) \quad \text{and} \quad \phi(+\infty) = \lim_{x \to \infty} \phi(x).$$

THEOREM 5.5   Let ϕ be any representative of $\mathcal{L}$. Then for any real number x_0,

$$\phi(x_0) - \phi(-\infty) \leqq \left\{ \sum_{k=0}^{\infty} p_k^2(x_0) \right\}^{-1} \quad \text{if } -\infty < x_0 \leqq \xi_1$$

$$\phi(+\infty) - \phi(x_0) \leqq \left\{ \sum_{k=0}^{\infty} p_k^2(x_0) \right\}^{-1} \quad \text{if } \eta_1 \leqq x_0 < +\infty.$$

Proof   Suppose that $-\infty < x_0 \leqq \xi_1$ and let $q(x)$ be a real quasi-orthogonal polynomial of order and degree $n + 1$ that has x_0 as a zero. Again defining $\rho(x)$ by (5.5), we have

$$\Lambda_{n0} = \mathcal{L}[\rho^2(x)] = \int_{-\infty}^{\infty} \rho^2(x) \, d\phi(x) \geqq \int_{-\infty}^{x_0} \rho^2(x) \, d\phi(x).$$

According to Theorem 5.2, $q(x)$ has no zero smaller than ξ_1 other than x_0. It follows that $\rho^2(x) > \rho^2(x_0) = 1$ for $x < x_0$, hence

$$\Lambda_{n0} \geqq \int_{-\infty}^{x_0} d\phi(x) = \phi(x_0) - \phi(-\infty).$$

By the preceding corollary, the desired conclusion now follows for $x \leqq \xi_1$. The remaining case is proved in the same way.   ∎

THEOREM 5.6   If $[\xi_1, \eta_1]$ is bounded, $\mathcal{L}$ is determinate.

Proof   It will be proved in Chapter IV (Theorem 2.5) that if $[\xi_1, \eta_1]$ is bounded, then

$$\lim_{n \to \infty} p_n^2(x) = \infty \quad \text{for } x \notin [\xi_1, \eta_1].$$

Assuming this result for now, it then follows from Theorem 5.5 that the spectrum of every representative of $\mathcal{L}$ is a subset of $[\xi_1, \eta_1]$. The proof of our theorem will now be completed by establishing the following general result.

THEOREM 5.7   Let $[a, b]$ be a compact interval and let ϕ_1 and ϕ_2 be functions of bounded variation on $[a, b]$ such that

$$\int_a^b x^n \, d\phi_1(x) = \int_a^b x^n \, d\phi_2(x), \qquad n = 0, 1, 2, \ldots$$

Then there exists a constant C such that $\phi_1(x) - \phi_2(x) = C$ at all $x \in [a, b]$ at which both are continuous.

Proof Let

$$\phi(x) = \phi_1(x) - \phi_2(x)$$

so that ϕ is of bounded variation on $[a, b]$ and

$$\int_a^b x^n \, d\phi(x) = 0,$$

hence

$$\int_a^b \pi(x) \, d\phi(x) = 0 \tag{5.6}$$

for every polynomial $\pi(x)$.

Now if f is any continuous function on $[a, b]$, then by the Weierstrass approximation theorem f can be uniformly approximated by polynomials. Because of (5.6) it then follows that

$$\int_a^b f(x) \, d\phi(x) = 0.$$

In particular, taking $f(x) = 1$ shows that $\phi(b) - \phi(a) = 0$.

Now for any t, $a < t < b$, which is a point of continuity of ϕ, define

$$f(x) = \begin{cases} x \text{ for } a \leq x \leq t \\ t \text{ for } t < x \leq b. \end{cases}$$

Then f is continuous on $[a, b]$ and we have

$$0 = \int_a^b f \, d\phi(x) = \int_a^t x \, d\phi(x) + \int_t^b t \, d\phi(x)$$

$$= t\phi(t) - a\phi(a) - \int_a^t \phi(x) \, dx + t\phi(b) - t\phi(t)$$

$$= (t - a)\phi(a) - \int_a^t \phi(x) \, dx.$$

Thus

$$\Phi(t) \equiv \int_a^t \phi(x)\,dx = (t-a)\phi(a),$$

and since ϕ is continuous at t, $\Phi'(t)$ exists and we have

$$\Phi'(t) = \phi(t) = \phi(a). \quad \blacksquare$$

The preceding completes the principal reason for this section. However, we observe that we have now developed sufficient machinery to validate partially a remark made in I-§5 to the effect that determinacy of $\mathcal{L}$ is equivalent to the existence of a smallest closed supporting set for $\mathcal{L}$.

THEOREM **5.8**   Let $\mathcal{L}$ be determinate and let ψ denote its substantially unique representative. Then $\mathfrak{S}(\psi)$ is a subset of every closed supporting set for $\mathcal{L}$.

Proof   Assume there is a closed supporting set F and a spectral point s such that $s \notin F$. Since F is closed, there is an open interval (α, β) such that

$$\alpha < s < \beta, \qquad (\alpha, \beta) \cap F = \varnothing. \tag{5.7}$$

Let $Q_n(x) = Q_n(x, \beta)$ denote the quasi-orthogonal polynomial of order $n+1$ which has a zero at β. If $Q_n(x)$ has a zero $z_n < \beta$, then a simple modification in the proof of Theorem 4.1 shows that $(z_n, \beta) \cap F \neq \varnothing$ (see Ex. 4.5). It then follows from (5.7) that $z_n < \alpha$. Thus in every case, $Q_n(x)$ has no zero in (α, β). Therefore the step function ϕ_n constructed from $Q_n(x)$ as in (5.4) is constant in (α, β).

We now consider the sequence $\{\phi_n\}_{n=1}^{\infty}$ and conclude that there is a subsequence which converges to a representative ϕ of $\mathcal{L}$ and that ϕ is constant on (α, β). Thus $\mathcal{L}$ would be indeterminate. $\blacksquare$

The converse to the above theorem can be deduced from a theorem (whose proof is beyond the level of this text) that if $\mathcal{L}$ is indeterminate, then for every closed set F there is a representative of $\mathcal{L}$ whose spectrum has F as its derived set.

Exercises

5.1   Complete the proof of Theorem 5.1.

5.2   Prove that the zeros of any two linearly independent quasi-orthogonal polynomials of the same order mutually separate each other.

5.3 Write out the formal proof of Theorem 5.3.

5.4 How should Theorem 5.4 be modified if x_0 is allowed to be a zero of $P_n(x)$?

5.5 Assume the notation of Theorem 5.3. Let $\psi(x) = -\phi(-x)$ and $r(x) = q(-x)$. Then ψ is a distribution function and $r(x)$ is a quasi-orthogonal polynomial of order $n + 1$ corresponding to the functional $\mathcal{L}^-$ determined by ψ. The zeros of $r(x)$ are $-y_i (i = 0, 1, \ldots, n)$. Show that in the quadrature formula corresponding to $\mathcal{L}^-$, the quadrature coefficient corresponding to $-y_k$ is $B_{n,n-k}$.

5.6 (The *Tchebichef inequalities*) Assume the notation of Theorem 5.3.
  (a) Show that there exists a non-negative polynomial $\pi(x)$ of degree $2n$ such that for $0 \leqq p < n$,
    (i) $\pi(y_i) = 1$ for $0 \leqq i \leqq p$;
    (ii) $\pi(y_j) = 0$ for $p < j \leqq n$;
    (iii) $\pi(x) \geqq 1$ for $x \leqq y_p$;
    (iv) $\pi(x)$ is decreasing for $y_p \leqq x \leqq y_{p+1}$. (fig. 1)
  (b) Use $\pi(x)$ to prove that if ϕ is any representative of $\mathcal{L}$,

$$\int_{-\infty}^{y_p+0} d\phi(x) < B_{n0} + B_{n1} + \cdots + B_{np}.$$

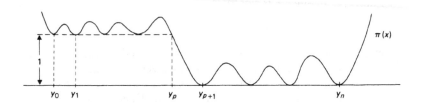

Figure 1

  (c) Use (b) and Ex. 5.5 to show that

$$\int_{y_{n-p}-0}^{\infty} d\phi(x) < B_{nn} + B_{n,n-1} + \cdots + B_{n,n-p},$$

  and then conclude that for $0 < p < n$,

$$\int_{-\infty}^{y_p+0} d\phi(x) < B_{n0} + B_{n1} + \cdots + B_{np} < \int_{-\infty}^{y_{p+1}-0} d\phi(x).$$

  (d) Prove finally that for $0 \leqq p \leqq n$,

$$\phi(y_p + 0) - \phi(y_p - 0) \leqq \phi_n(y_p + 0) - \phi_n(y_p - 0)$$

  with strict inequality holding for $p \neq 0, n$.

5.7 Prove that if $\Lambda_0 = \{\sum_{k=0}^{\infty} p_k^2(x_0)\}^{-1} > 0$, then there is a representative of $\mathcal{L}$

having a jump equal to Λ_0 at x_0 and no other representative of $\mathfrak{L}$ has a larger jump at x_0.

5.8 Let $\{P_n(x)\}$ be a monic OPS with respect to a quasi-definite $\mathfrak{L}$. Let $\{Q_n(x)\}$ be a sequence of monic polynomials such that $Q_0(x) = 1$ and $Q_n(x)$ is a quasi-orthogonal polynomial of order and degree n relative to $\mathfrak{L}$. Prove that if $Q_1(x) \neq P_1(x)$ and $\{Q_n(x)\}$ is an OPS, then $\{P_n(x)\}$ is a monic KPS for $\{Q_n(x)\}$.

6 The Classical Moment Problems

We will discuss rather briefly the classical Stieltjes and Hamburger moment problems for it is here that the origins of the general theory of orthogonal polynomials are to be found. (Particular examples of orthogonal polynomials—the so called "classical orthogonal polynomials"—had been studied previously.)

The subject had its beginnings in the investigations of P. L. Tchebichef (beginning in 1874) and his pupil A. Markov on the "theory of limiting values of integrals." In this work, Tchebichef introduced many important new ideas, among them a general concept of orthogonal polynomials. We refer to M. G. Krein [1] for a survey of this work and the subsequently developed theory and for many interesting historical remarks. See also the recent book by Karlin and Studden [1].

Extending and generalizing a problem implicit in Tchebichef's investigations, T. J. Stieltjes posed and solved the following problem in 1894:

THE STIELTJES MOMENT PROBLEM   Given a real sequence $\{\mu_n\}_{n=0}^{\infty}$, find necessary and sufficient conditions in order that there exists a distribution function ψ with an infinite spectrum contained in $[0, \infty)$ such that

$$\int_0^{\infty} x^n \, d\psi(x) = \mu_n, \qquad n = 0, 1, 2, \ldots \tag{6.1}$$

In order to pose the problem in this generality, Stieltjes invented the Stieltjes integral in an eventful paper [2] in which he also explored thoroughly uniqueness questions and in the process introduced many new and important ideas into analysis.

In 1920-21, H. Hamburger made an important extension of the problem by allowing the spectrum of ψ to be in $(-\infty, \infty)$. That is, he asked for the existence of a distribution function ψ such that

$$\int_{-\infty}^{\infty} x^n \, d\psi(x) = \mu_n, \qquad n = 0, 1, 2, \ldots \tag{6.2}$$

We cannot go into the numerous contributions made by these authors and the many others (in particular, P. Nevanlinna, M. Riesz, T. Carleman,

F. Hausdorff, M. H. Stone) who further developed the theory by many different methods. Further historical comments and an indication of the many varied methods that have been brought to bear on the problem can be found together with a thorough discussion of the theory in the book by Shohat and Tamarkin [1].

Addressing only the simpler existence question for the Hamburger moment problem (6.2), we see in view of Theorem 3.1 that a necessary and sufficient condition that there be a solution is that the corresponding moment functional be positive-definite. Combining this observation with I-Theorem 3.4, we can then state Hamburger's criterion.

THEOREM 6.1   A necessary and sufficient condition that the Hamburger moment problem have a solution is that the determinants Δ_n given by I-(3.1) are all positive.

Turning next to the Stieltjes moment problem (6.1), we see that according to Theorem 3.2 a necessary and sufficient condition that there is a solution is that the corresponding moment functional is positive-definite on $[0, \infty]$.

Referring to I-Theorem 7.1, consider the corresponding functional $\mathcal{L}_0^*$. We see that if $\mathcal{L}$ is positive-definite on $[0, \infty]$, so is $\mathcal{L}_0^*$. Conversely, if $\mathcal{L}$ and $\mathcal{L}_0^*$ are both positive-definite, then $\kappa = 0 \leqq \xi_1$, hence $\mathcal{L}$ is positive-definite on $[0, \infty]$. That is, a necessary and sufficient condition for the Stieltjes moment problem to have a solution is that $\mathcal{L}$ and $\mathcal{L}_0^*$ are both positive-definite.

Now the moments of $\mathcal{L}_0^*$ are $\mu_n^* = \mu_{n+1}$ ($n \geqq 0$), hence introducing

$$\Delta_n^{(1)} = \det(\mu_{i+j+1}) = \begin{vmatrix} \mu_1 & \mu_2 & \cdots & \mu_{n+1} \\ \mu_2 & \mu_3 & \cdots & \mu_{n+2} \\ . & . & \cdots & . \\ . & . & \cdots & . \\ \mu_{n+1} & \mu_{n+2} & \cdots & \mu_{2n+1} \end{vmatrix}$$

we can state Stieltjes' criterion.

THEOREM 6.2   A necessary and sufficient condition that the Stieltjes moment problem have a solution is

$$\Delta_n > 0, \qquad \Delta_n^{(1)} > 0, \qquad n = 0, 1, 2, \ldots$$

When a moment problem has a solution and the corresponding moment functional is determinate, the moment problem is said to be *determined*. Otherwise, it is said to be an *indeterminate* moment problem.

According to Theorem 5.6, if the moment problem has a solution whose spectrum is a bounded set, then it will be a determined moment problem. From the point of view of orthogonal polynomials, this is a satisfactory result since this state of affairs can be determined immediately from the coefficients in the fundamental recurrence formula (see IV-Theorem 2.2). From the viewpoint of the problem of moments itself, this leaves something to be desired since what is needed is a criterion expressed directly in terms of the moments. Such a criterion was provided by F. Hausdorff who solved the "Hausdorff moment problem" in which the spectrum of ψ is required to lie in $[0, 1]$. As noted, Theorem 5.6 shows that a Hausdorff moment problem which is solvable is always determined.

As indicated previously, we will not take up the general question of determinacy here. In addition to the previously mentioned work by Shohat and Tamarkin, an elegant exposition can be found in the recent book by N. I. Ahiezer [1]. We will at this time give a classical example due to Stieltjes of an indeterminate moment problem.

In both of the definite integrals,

$$\int_{-\infty}^{\infty} e^{-u^2}\,du = \sqrt{\pi} \quad \text{and} \quad \int_{-\infty}^{\infty} e^{-u^2}\sin(2\pi u)\,du = 0,$$

make the substitution, $u = \log x - (n + 1)/2$, to obtain

$$\int_0^{\infty} x^n \exp(-\log^2 x)\,dx = \sqrt{\pi}\,\exp[(n + 1)^2/4]$$

$$\int_0^{\infty} x^n \sin(2\pi \log x)\exp(-\log^2 x)\,dx = 0.$$

Thus we have for arbitrary C,

$$\int_0^{\infty} x^n[1 + C\sin(2\pi \log x)]\exp(-\log^2 x)\,dx = \sqrt{\pi}\,\exp[(n + 1)^2/4].$$

Since for $|C| < 1$, the functions

$$\psi(C; x) = \int_0^x [1 + C\sin(2\pi \log t)]\exp(-\log^2 t)\,dt$$

are clearly non-decreasing, this shows that there are infinitely many distribution functions with spectra in $[0, \infty)$ having the moments, $\mu_n = \sqrt{\pi}\,\exp[(n + 1)^2/4]$. Thus the corresponding Stieltjes moment problem is indeterminate (as is also the corresponding Hamburger moment problem). The corresponding orthogonal polynomials are called *Stieltjes-Wigert* polynomials and are discussed in VI-§3 (see also I-Ex. 9.11).

We conclude our discussion of moment problems with a theorem of R.P. Boas which provides a representation theorem for moment functionals in general.

THEOREM **6.3**   Let $\{\mu_n\}$ be an arbitrary sequence of real numbers. Then there is a function ϕ of bounded variation such that

$$\int_{-\infty}^{\infty} x^n \, d\phi(x) = \mu_n, \qquad n = 0, 1, 2, \ldots$$

Proof   Write

$$\mu_0 = \mu_{10} - \mu_{20}$$

where $\mu_{10} > 0$ and $\mu_{20} > 0$.

Now assume $\mu_{i0}, \mu_{i1}, \ldots, \mu_{i,2n}$ $(i = 1, 2)$ have been determined such that
(i)

$$\mu_k = \mu_{1k} - \mu_{2k}, \qquad k = 0, 1, 2, \ldots, 2n,$$

(ii)

$$\Delta_{pk} = \det(\mu_{p,i+j})_{i,j=0}^{k} > 0 \qquad \text{for } p = 1, 2; \ k = 0, 1, \ldots, n.$$

Choose any $\mu_{1,2n+1}$ and $\mu_{2,2n+1}$ such that

$$\mu_{2n+1} = \mu_{1,2n+1} - \mu_{2,2n+1}.$$

Now for any such choices, $\Delta_{p,n+1}$ can be expanded by the elements of its last row to obtain

$$\Delta_{p,n+1} = \mu_{p,2n+2} + f(\mu_{p,0}, \mu_{p,1}, \ldots, \mu_{p,2n+1}), \qquad p = 1, 2,$$

where f is independent of $\mu_{p,2n+2}$. It follows that $\Delta_{p,n+1} > 0$ for $\mu_{p,2n+2}$ sufficiently large.

Since it is always possible to choose arbitrarily large $\mu_{p,2n+2}$ such that (i) holds for $k = 2n + 2$, this means that it is always possible to choose $\mu_{p,2n+2}$ such that both (i) and (ii) are valid for $k = 2n + 2$.

By induction it follows that there exist moment sequences $\{\mu_{1,n}\}$ and $\{\mu_{2,n}\}$ such that (i) holds for all k and the corresponding Hamburger moment problems have solutions, ψ_1 and ψ_2, respectively. Thus the original moment problem has a solution $\phi = \psi_1 - \psi_2$ which is a function of bounded variation.   ■

It thus follows that every moment functional $\mathcal{L}$ with real moments can be represented by a Stieltjes integral whose integrator is a function of bounded variation. If instead, $\mathcal{L}$ has complex moments $\mu_n = \alpha_n + i\beta_n$, then $\mathcal{L}$ can be represented by the complex-valued function of bounded variation, $\phi = \phi_1 + i\phi_2$, where ϕ_1 and ϕ_2 are the functions of bounded variation whose moment sequences are $\{\alpha_n\}$ and $\{\beta_n\}$.

It should be noted that the moment problem of Theorem 6.3 is always indeterminate since we can always add to a solution of the moment problem an arbitrary function of bounded variation all of whose moments are 0. A non-trivial example of such a function appears in our earlier example of an indeterminate Stieltjes moment problem.

Finally, it may be worthwhile to restate I-Theorem 4.4 in the light of our representation theorems.

THEOREM **6.4**   Let $\{c_n\}$ and $\{\lambda_n\}$ be arbitrary complex sequences and let $\{P_n(x)\}$ be defined by

$$P_n(x) = (x - c_n)P_{n-1}(x) - \lambda_n P_{n-2}(x), \qquad n = 1, 2, 3, \ldots$$

$$P_{-1}(x) = 0, \qquad P_0(x) = 1.$$

Then there is a function ϕ of bounded variation on $(-\infty, \infty)$ such that

$$\int_{-\infty}^{\infty} P_m(x)P_n(x)\,d\phi(x) = \lambda_1\lambda_2\cdots\lambda_{n+1}\delta_{mn}, \qquad m, n = 0, 1, 2, \ldots$$

ϕ can be chosen to be real-valued if and only if $\{c_n\}$ and $\{\lambda_n\}$ are real sequences. ϕ can be chosen to be non-decreasing with an infinite spectrum if and only if c_n is real and $\lambda_n > 0$ for all n.

Exercises

6.1   Let $\{\mu_n\}$ be an indeterminate Stieltjes moment sequence (i.e., the corresponding Stieltjes moment problem has a solution but is indeterminate). Write $\mu_{2n}^* = \mu_n$, $\mu_{2n+1}^* = 0$ $(n \geq 0)$. Show that $\{\mu_n^*\}$ is an indeterminate Hamburger moment sequence.

6.2   Give an example of a non-negative, continuous function w such that

$$0 < \int_{-\infty}^{\infty} x^{2n}w(x)\,dx < \infty, \qquad \int_{-\infty}^{\infty} x^{2n+1}w(x)\,dx = 0,$$

but w is not an even function.

6.3   Carleman proved that the Hamburger moment problem is determined if

$$\sum_{n=1}^{\infty} \mu_{2n}^{-1/(2n)} = \infty.$$

Assuming this, prove that the Stieltjes moment problem is determined if

$$\sum_{n=1}^{\infty} \mu_n^{-1/(2n)} = \infty.$$

6.4 Let $\{\mu_n\}$ be a Stieltjes moment sequence. Prove that $\mu_n^2 \leq \mu_{n+1}\mu_{n-1}$ $(n \geq 1)$.

6.5 Prove that a bounded Stieltjes moment sequence is a Hausdorff moment sequence (spectrum of a solution lies in $[0,1]$).

6.6 Let $\{\mu_n\}$ be a Stieltjes moment sequence and suppose the series

$$\sum_{n=1}^{\infty} \frac{(-4\pi^2)^n \mu_{2n}}{(2n)!}$$

converges absolutely. Prove that $\{\mu_n\}$ is the moment sequence of a distribution function whose spectrum is a set of non-negative integers if and only if the sum of the series is 0. (Wintner [1]).

6.7 For any sequence $\{a_n\}_{n=0}^{\infty}$, define

$$\Delta^0 a_n = a_n, \qquad \Delta^{k+1} a_n = \Delta^k a_n - \Delta^k a_{n+1} \qquad (k, n = 0, 1, 2, \ldots).$$

(i) Prove that

$$\Delta^k a_n = \sum_{i=0}^{k} \binom{k}{i}(-1)^i a_{n+i}.$$

(ii) $\{a_n\}$ is called *completely monotonic* if $\Delta^k a_n \geq 0$ for all k, n. Prove that if ψ is a non-decreasing function whose spectrum is in $[0,1]$, then its moment sequence is completely monotonic. (The converse is also true but considerably more difficult.)

CHAPTER III

Continued Fractions and Chain Sequences

Continued fractions play a fundamental role in most investigations into the classical moment problems. Orthogonal polynomials arise in a natural way in the analysis of certain types of continued fractions associated with moment problems and this fact can be made the basis of the development of a theory of orthogonal polynomials.

We will not take this viewpoint but will consider continued fractions just enough to indicate the relation between continued fractions and orthogonal polynomials and to obtain certain results that will be needed in our further study of the properties of orthogonal polynomials.

1 Basic Concepts

Let $\{a_n\}_{n=1}^{\infty}$ and $\{b_n\}_{n=0}^{\infty}$ be arbitrary sequences of complex numbers and write

$$
\begin{aligned}
C_0 &= b_0 \\
C_1 &= b_0 + \frac{a_1}{b_1} \\
C_2 &= b_0 + \cfrac{a_1}{b_1 + \cfrac{a_2}{b_2}} \\
&\cdots\cdots \\
C_n &= b_0 + \cfrac{a_1}{b_1 + \cfrac{a_2}{b_2 + \cfrac{\ddots}{\ddots\; + \cfrac{a_n}{b_n}}}}
\end{aligned}
\tag{1.1}
$$

C_n is called the nth *convergent* (or *approximant*) of the (infinite) *continued fraction*

77

$$b_0 + \cfrac{a_1}{b_1 + \cfrac{a_2}{b_2 + \cfrac{\ddots}{\quad + \cfrac{a_n}{b_n + \ddots}}}} \tag{1.2}$$

The above "definition" involves a logical lacuna analogous with what is a frequent occurence in elementary treatments of infinite series. Namely it does not define what a continued fraction *is*. We might therefore define a continued fraction to be an ordered triple $(\{a_n\}, \{b_n\}, \{C_n\})$, where $\{C_n\}$ is defined by (1.1). Alternately, we could define a continued fraction as a formal expression of the form (1.2). In any case, the important matter is the behavior of the sequence of convergents, $\{C_n\}$.

Since the expressions (1.1) and (1.2) are rather space consuming, we will adopt the notations

$$C_n = b_0 + \frac{a_1|}{|b_1} + \frac{a_2|}{|b_2} + \cdots + \frac{a_n|}{|b_n} \tag{1.3}$$

$$b_0 + \frac{a_1|}{|b_1} + \frac{a_2|}{|b_2} + \cdots + \frac{a_n|}{|b_n} + \cdots \tag{1.4}$$

in their place. We also adopt the convention that if $a_i = -d_i$, then the above will be modified by writing

$$-\frac{d_i|}{|b_i} \quad \text{in place of} \quad +\frac{-d_i|}{|b_i}.$$

Other notation in more or less common use include

$$\left\{ b_0; \frac{a_n}{b_n} \right\}_{n=1}^{\infty} \quad \text{and} \quad b_0 + K_{n=1}^{\infty} \frac{a_n}{b_n}$$

and, most frequently,

$$b_0 + \frac{a_1}{b_1} + \frac{a_2}{b_2} + \cdots + \frac{a_n}{b_n} + \cdots.$$

It is of course possible that certain of the convergents will be undefined. For example, C_2 is meaningless if $b_1 b_2 = -a_2$. Nevertheless we may still wish to consider the corresponding continued fraction, hence we make the following definition.

DEFINITION 1.1   The continued fraction (1.4) is said to *converge* to the value K (finite) provided at most finitely many C_n are undefined and

$$\lim_{n \to \infty} C_n = K.$$

Otherwise, the continued fraction is said to *diverge*.

If the continued fraction converges to K, we will write

$$b_0 + \frac{a_1|}{|b_1} + \frac{a_2|}{|b_2} + \cdots = K.$$

This involves the same ambiguity that occurs with infinite series when we write $\sum a_n = S$ so that $\sum a_n$ denotes both the series and its sum.

Exercises

1.1   Two continued fractions are called *equivalent* if they have the same sequence of nth convergents (in the sense that if C_n and C'_n denote their respective nth convergents, then either $C_n = C'_n$ or both are undefined). Show that the following continued fractions are equivalent:

(i) $b_0 + \dfrac{a_1|}{|b_1} + \dfrac{a_2|}{|b_2} + \dfrac{a_3|}{|b_3} + \cdots$

(ii) $b_0 + \dfrac{k_1 a_1|}{|k_1 b_1} + \dfrac{k_1 k_2 a_2|}{|\ k_2 b_2} + \dfrac{k_2 k_3 a_3|}{|\ k_3 b_3} + \cdots,$

where $k_i \neq 0$.

1.2   Show that if a continued fraction is equivalent to the continued fraction in Ex. 1.1 (i), and if $a_i \neq 0$ for all i, then it has the form of the continued fraction in Ex. 1.1 (ii) for suitably chosen k_i.

1.3   Show that the continued fraction

$$\frac{\lambda_1 \ |}{|x - c_1} - \frac{\lambda_2 \ |}{|x - c_2} - \frac{\lambda_3 \ |}{|x - c_3} - \cdots$$

is equivalent to

$$\frac{\alpha_0(x)|}{|\ 1} - \frac{\alpha_1(x)|}{|\ 1} - \frac{\alpha_2(x)|}{|\ 1} - \cdots$$

where

$$\alpha_0(x) = \frac{\lambda_1}{x - c_1}, \qquad \alpha_n(x) = \frac{\lambda_{n+1}}{(c_n - x)(c_{n+1} - x)} \qquad (n \geq 1).$$

1.4 Prove that if

$$b_0 + \frac{a_1|}{|b_1} + \frac{a_2|}{|b_2} + \frac{a_3|}{|b_3} + \cdots = K \neq 0,$$

then

$$b_{-1} + \frac{a_0|}{|b_0} + \frac{a_1|}{|b_1} + \frac{a_2|}{|b_2} + \frac{a_3|}{|b_3} + \cdots = b_{-1} + \frac{a_0}{K}.$$

1.5 Show that if the continued fraction

$$1 + \frac{1|}{|1} + \frac{1|}{|1} + \frac{1|}{|1} + \cdots$$

converges, then its value is $(1 + \sqrt{5})/2$.

2 The Fundamental Recurrence Formulas

Referring to (1.1), we can write

$$C_n = \frac{A_n}{B_n}, \qquad n = 0, 1, 2, \ldots$$

where

$$A_0 = b_0, \qquad B_0 = 1,$$
$$A_1 = b_0 b_1 + a_1, \qquad B_1 = b_1,$$
$$A_2 = b_0 b_1 b_2 + b_0 a_2 + a_1 b_2, \qquad B_2 = b_1 b_2 + a_2,$$

and in general, A_n and B_n are polynomials in a_i, b_j.
   Now we can write

$$A_1 = b_1 A_0 + a_1 A_{-1} \qquad \text{where } A_{-1} = 1,$$
$$B_1 = b_1 B_0 + a_1 B_{-1} \qquad \text{where } B_{-1} = 0.$$

Assume that for some $n \geq 1$,

$$A_n = b_n A_{n-1} + a_n A_{n-2}, \qquad A_{-1} = 1,$$
$$B_n = b_n B_{n-1} + a_n B_{n-2} \qquad B_{-1} = 0. \tag{2.1}$$

Then since C_{n+1} can be obtained from C_n by replacing b_n by $b_n + (a_{n+1})/b_{n+1}$, we can write

$$C_{n+1} = \frac{A_{n+1}}{B_{n+1}} = \frac{A_n^*}{B_n^*}$$

where by (2.1)

$$A_n^* = \left(b_n + \frac{a_{n+1}}{b_{n+1}}\right)A_{n-1} + a_n A_{n-2}$$
$$= b_{n+1}^{-1}[b_{n+1}(b_n A_{n-1} + a_n A_{n-2}) + a_{n+1}A_{n-1}]$$
$$= b_{n+1}^{-1}[b_{n+1}A_n + a_{n+1}A_{n-1}].$$

In exactly the same way, we obviously obtain

$$B_n^* = b_{n+1}^{-1}[b_{n+1}B_n + a_{n+1}B_{n-1}],$$

so that, whenever it is defined,

$$C_{n+1} = \frac{b_{n+1}A_n + a_{n+1}A_{n-1}}{b_{n+1}B_n + a_{n+1}B_{n-1}}.$$

It now follows by induction that if $b_i \neq 0$ for $i \geq 1$, then the formulas (2.1), first established by Wallis in 1655, are valid for $n \geq 1$.

A_n and B_n are called the nth *partial numerator* and nth *partial denominator*, respectively.

If the first equation in (2.1) is multiplied by B_{n-1} and the second by A_{n-1}, then exactly as in the derivation of the Christoffel-Darboux identity (I-(4.9)), we obtain

$$A_n B_{n-1} - B_n A_{n-1} = -a_n[A_{n-1}B_{n-2} - B_{n-1}A_{n-2}],$$

whence

$$A_n B_{n-1} - B_n A_{n-1} = (-1)^{n+1}a_1 a_2 \cdots a_n, \qquad n \geq 1. \qquad (2.2)$$

The latter can be written

$$\frac{A_n}{B_n} - \frac{A_{n-1}}{B_{n-1}} = \frac{(-1)^{n+1}a_1 a_2 \cdots a_n}{B_{n-1}B_n}.$$

Recalling that $A_0/B_0 = b_0$, we thus have the important formula

$$\frac{A_n}{B_n} = b_0 + \sum_{k=1}^{n} \frac{(-1)^{k+1}a_1 a_2 \cdots a_k}{B_{k-1}B_k} \qquad (2.3)$$

provided $b_i \neq 0$, $B_i \neq 0$ $(1 \leq i \leq n)$.

Exercises

2.1 Let $a_{i+1} \geq 1$, $b_i \geq 1$ ($i \geq 0$). Show that $A_n \geq f_{n+1}$, $B_n \geq f_n$ where $\{f_n\}$ is the Fibonacci sequence: $f_0 = 1, f_1 = 1, f_n = f_{n-1} + f_{n-2}$ ($n \geq 2$).

2.2 Prove that if $b_i \geq 1$ for $i \geq 1$, then the continued fraction

$$b_0 + \frac{1|}{|b_1} + \frac{1|}{|b_2} + \frac{1|}{|b_3} + \cdots$$

converges.

2.3 Let the b_i be positive integers. Prove that the continued fraction in Ex. 2.2 converges to an irrational number. [Hint: Show that if $\{A_n/B_n\}$ converges to a rational limit $x = p/q$, then $|xB_n - A_n| < 1/q$ for n sufficiently large.]

2.4 Let the b_i be positive integers so the continued fraction in Ex. 2.2 converges to a limit x.
   (a) Show that $b_0 = [x]$, where $[x]$ denotes the greatest integer not exceeding x.
   (b) Prove that two distinct continued fractions having the form in Ex. 2.2 with the b_i positive integers converge to distinct limits.

2.5 Let $a_i > 0$, $b_i > 0$. Prove that

$$\left\{ \frac{a_1 a_2 \cdots a_n}{B_{n-1} B_n} \right\}$$

is a decreasing sequence, hence prove that (1.4) converges if

$$\lim_{n \to \infty} \frac{a_1 a_2 \cdots a_n b_n}{(b_1 b_2 \cdots b_n)^2} = 0.$$

2.6 Let $a_i > 0$, $b_i > 0$. Prove that $\{C_{2n}\}$, the subsequence of convergents of (1.4) of even order, is increasing and convergent. State and prove a corresponding result for $\{C_{2n+1}\}$.

2.7 Referring to the continued fraction in Ex. 2.2,
   (a) Show that $|A_n| \leq M \prod_{k=1}^{n} (1 + |b_k|)$, $M = \max(1, |b_0|)$.
   (b) Prove that if $\sum_{k=1}^{\infty} |b_{2k}|$ converges, then $\{A_{2n}\}$ converges.
   (c) Prove that if $\sum_{k=1}^{\infty} |b_k|$ converges, then all four sequences, $\{A_{2n}\}$, $\{A_{2n+1}\}$, $\{B_{2n}\}$, $\{B_{2n+1}\}$, converge.
   (d) Conclude that if $\sum_{k=1}^{\infty} |b_k|$ converges, the continued fraction in Ex. 2.2 diverges.

3 A Convergence Theorem

We obtain a convergence theorem for a continued fraction of a very special type which will be used in section 6.

LEMMA 3.1 Let $m_0 = 0$ Then the nth partial denominator of the continued fraction

$$1 - \frac{1|}{|1} - \frac{(1-m_0)m_1|}{|1} - \frac{(1-m_1)m_2|}{|1} - \cdots$$

is

$$B_n = (1-m_0)(1-m_1)\cdots(1-m_{n-1}), \qquad n \geq 1. \tag{3.1}$$

Proof Using the Wallis formulas (2.1),

$$B_{N+1} = B_N - (1 - m_{N-1})m_N B_{N-1}$$
$$= (1-m_0)(1-m_1)\cdots(1-m_{N-1})(1-m_N)$$

provided (3.1) holds for $n \leq N$. Since $B_1 = 1 = 1 - m_0$, (3.1) follows by induction. ∎

LEMMA 3.2 Let $\alpha_n = (1 - m_{n-1})m_n$ where $m_0 = 0$, $0 < m_n < 1$ ($n = 1, 2, 3, \ldots$). Then the continued fraction

$$1 - \frac{\alpha_1|}{|1} - \frac{\alpha_2|}{|1} - \frac{\alpha_3|}{|1} - \cdots$$

converges to $(1 + L)^{-1}$ where

$$L = \sum_{n=1}^{\infty} \frac{m_1 m_2 \cdots m_n}{(1-m_1)(1-m_2)\cdots(1-m_n)}.$$

Proof Let A_n/B_n denote the nth convergent of the given continued fraction and let

$$\frac{A_{n+1}^*}{B_{n+1}^*} = 1 - \frac{1}{\dfrac{A_n}{B_n}} = 1 - \frac{1|}{|1} - \frac{\alpha_1|}{|1} - \cdots - \frac{\alpha_n|}{|1}.$$

By Lemma 3.1, $B_{n+1}^* = (1 - m_0)\cdots(1 - m_n) > 0$, hence using (2.3)

$$\frac{A_{n+1}^*}{B_{n+1}^*} = 1 - \sum_{k=0}^{n} \frac{\alpha_0 \alpha_1 \cdots \alpha_k}{B_k^* B_{k+1}^*} \quad (\alpha_0 = 1)$$

$$= - \sum_{k=1}^{n} \frac{\alpha_1 \alpha_2 \cdots \alpha_k}{(1-m_0)^2 \cdots (1-m_{k-1})^2(1-m_k)}$$

$$= - \sum_{k=1}^{n} \frac{m_1 m_2 \cdots m_k}{(1-m_1)(1-m_2)\cdots(1-m_k)}.$$

Thus

$$\lim_{n \to \infty} \frac{A_n}{B_n} = \lim_{n \to \infty} \frac{1}{1 - \frac{A_{n+1}^*}{B_{n+1}^*}} = \frac{1}{1 + L}. \qquad \blacksquare$$

THEOREM 3.1   Let $\beta_n = (1 - g_{n-1})g_n$ where $0 \leqq g_0 < 1$ and $0 < g_n < 1$ for $n \geqq 1$. Then

$$1 - \frac{\beta_1|}{|1} - \frac{\beta_2|}{|1} - \cdots - \frac{\beta_n|}{|1} - \cdots = g_0 + \frac{1 - g_0}{1 + G} \qquad (3.2)$$

where

$$G = \sum_{n=1}^{\infty} \frac{g_1 g_2 \cdots g_n}{(1 - g_1)(1 - g_2) \cdots (1 - g_n)}.$$

Proof   The case $g_0 = 0$ is given by Lemma 3.2 so assume $g_0 > 0$. Let C_n denote the nth convergent of (3.2) and let C_n^* denote the nth convergent of the continued fraction

$$1 - \frac{g_0|}{|1} - \frac{\beta_1|}{|1} - \frac{\beta_2|}{|1} - \cdots$$

According to Lemma 3.2, C_n^* converges to $(1 + K)^{-1}$ where

$$K = \sum_{n=1}^{\infty} \frac{g_0 g_1 \cdots g_{n-1}}{(1 - g_0)(1 - g_1) \cdots (1 - g_{n-1})} = \frac{g_0}{1 - g_0}(1 + G).$$

Since $C_{n+1}^* = 1 - g_0 C_n^{-1}$, it follows that $\{C_n\}$ converges to $g_0(1 + K^{-1})$ and this yields (3.2). $\blacksquare$

Exercises

3.1   Use Theorem 3.1 to show that

$$1 - \frac{a|}{|1} - \frac{a|}{|1} - \frac{a|}{|1} - \cdots$$

converges for $0 < 4a < 1$ to $(1 + \sqrt{1 - 4a})/2$.

3.2   Put $\beta_1 = 1/2$, $\beta_n = (n - 1)/n(n + 1)$ $(n \geqq 2)$ in (3.2). Conclude that

$$1 - \frac{1|}{|2} - \frac{1|}{|3} - \frac{2|}{|4} - \frac{3|}{|5} - \frac{4|}{|6} - \cdots = \frac{1}{e}.$$

3.3 Show that

$$1 - \frac{1|}{|5} - \frac{2.4|}{|8} - \frac{2.9|}{|11} - \cdots - \frac{2n^2|}{|3n + 2} - \cdots = \frac{1}{2\ln 2}.$$

3.4 Show that

$$1 - \frac{1|}{|2} - \frac{1|}{|5} - \frac{2^2|}{|10} - \cdots - \frac{(n-1)^2|}{|n^2 + 1} - \cdots = \frac{1}{J_0(2i)},$$

where $J_0(x)$ denotes the Bessel function of order 0.

4 Jacobi Fractions and Orthogonal Polynomials

The Wallis formulas (2.1) lead directly to the connection between orthogonal polynomials and continued fractions. For if we take in (1.4)

$$b_0 = 0, \quad a_1 = \lambda_1 \neq 0, \quad a_{n+1} = -\lambda_{n+1} \neq 0, \quad b_n = x - c_n, \quad n \geq 1,$$

we then have the continued fraction

$$\frac{\lambda_1 \ |}{|x - c_1} - \frac{\lambda_2 \ |}{|x - c_2} - \frac{\lambda_3 \ |}{|x - c_3} - \cdots \qquad (4.1)$$

whose nth partial denominators, $B_n = P_n(x)$, satisfy

$$P_n(x) = (x - c_n)P_{n-1}(x) - \lambda_n P_{n-2}(x), \quad n = 1, 2, 3, \ldots$$
$$P_{-1}(x) = 0, \quad P_0(x) = 1. \qquad (4.2)$$

It thus follows from I-Theorem 4.4 that the denominators of (4.1) form a monic OPS with respect to a quasi-definite moment functional $\mathcal{L}$, which is positive-definite if the c_n are real and the λ_n are positive. The continued fraction (4.1) is called a *Jacobi* type continued fraction or simply a *J-fraction* because of its relation to certain matrices known as *Jacobi matrices* (see I-Ex. 5.7).

Returning to the Wallis formulas, we note that the partial numerators, $A_n = A_n(x)$, satisfy the recurrence

$$A_n(x) = (x - c_n)A_{n-1}(x) - \lambda_n A_{n-2}(x), \qquad n = 2, 3, \ldots$$
$$A_{-1}(x) = 1, \qquad A_0(x) = 0, \qquad A_1(x) = \lambda_1.$$

It is easy to verify by induction that $\lambda_1^{-1} A_n(x)$ is a monic polynomial of degree $n - 1$ which is independent of λ_1. Hence we write

$$P_n^{(1)}(x) = \lambda_1^{-1} A_{n+1}(x), \qquad n \geqq -1.$$

Then $P_n^{(1)}(x)$ is a monic polynomial of degree n independent of λ_1 which satisfies the recurrence

$$P_n^{(1)}(x) = (x - c_{n+1}) P_{n-1}^{(1)}(x) - \lambda_{n+1} P_{n-2}^{(1)}(x), \qquad n = 1, 2, 3, \ldots$$
$$P_{-1}^{(1)}(x) = 0, \qquad P_0^{(1)}(x) = 1. \tag{4.3}$$

Thus $\{P_n^{(1)}(x)\}$ is a monic OPS with respect to a quasi-definite moment functional $\mathcal{L}^{(1)}$, which is positive-definite if the c_n are real and the λ_n are positive for $n \geqq 2$.

DEFINITION 4.1   The polynomials $P_n^{(1)}(x)$ are called the monic *numerator polynomials* corresponding to $\{P_n(x)\}$. The term monic *numerator OPS* will also be used.

The name *associated polynomials* is frequently used in the literature in place of numerator polynomials.

From (2.2) we obtain the identity

$$P_{n+1}(x) P_{n-1}^{(1)}(x) - P_n^{(1)}(x) P_n(x) = -\lambda_2 \lambda_3 \cdots \lambda_{n+1}, \qquad n \geqq 1. \tag{4.4}$$

Now if c_n is real and $\lambda_n > 0$ so that $\mathcal{L}$ and $\mathcal{L}^{(1)}$ are both positive-definite, then $P_{n+1}(x)$ and $P_n^{(1)}(x)$ have real, simple zeros which are interlaced in the same manner as those of $P_{n+1}(x)$ and $P_n(x)$. Specifically, denote the zeros of $P_n^{(1)}(x)$ by

$$x_{n1}^{(1)} < x_{n2}^{(1)} < \cdots < x_{nn}^{(1)}.$$

We then have

THEOREM 4.1

$$x_{n+1,k} < x_{nk}^{(1)} < x_{n+1,k+1}, \qquad k = 1, 2, \ldots, n.$$

Proof (4.4) yields

$$P_n^{(1)}(x_{n+1,k})P_n(x_{n+1,k}) = \lambda_2\lambda_3\cdots\lambda_{n+1} > 0.$$

Invoking I-Theorem 5.3 and engaging in yet another argument concerning the variation in sign of $P_n(x_{n+1,k})$, we conclude that $P_n^{(1)}(x)$ has at least one zero on each open interval $(x_{n+1,k}, x_{n+1,k+1})$ for $1 \leqq k \leqq n$. ∎

COROLLARY   If $[\xi_1^{(1)}, \eta_1^{(1)}]$ denotes the true interval of orthogonality of $\wp^{(1)}$, then

(i) $[\xi_1^{(1)}, \eta_1^{(1)}] \subset [\xi_1, \eta_1]$;
(ii) If $\xi_1 < \xi_1^{(1)}$ (if $\eta_1 > \eta_1^{(1)}$), then for all n sufficiently large, $P_n(x)$ has exactly one zero on $(\xi_1, \xi_1^{(1)})$ (on $(\eta_1^{(1)}, \eta_1)$) .

Since $\{P_n^{(1)}(x)\}$ is an OPS in its own right, it has a corresponding monic numerator OPS. This suggests introducing the OPS $\{P_n^{(k)}(x)\}$ defined by

$$P_n^{(k)}(x) = (x - c_{n+k})P_{n-1}^{(k)}(x) - \lambda_{n+k}P_{n-2}^{(k)}(x), \qquad n = 1, 2, 3, \ldots$$
$$P_{-1}^{(k)}(x) = 0, \qquad P_0^{(k)}(x) = 1, \qquad k = 0, 1, 2, \ldots \tag{4.5}$$

Thus $\{P_n^{(0)}(x)\} = \{P_n(x)\}$ and $\{P_n^{(k)}(x)\}$ is the monic numerator OPS corresponding to $\{P_n^{(k-1)}(x)\}$.

Considering now the positive-definite case, we denote by $\{x_{ni}^{(k)}\}_{i=1}^n$ the zeros of $P_n^{(k)}(x)$ ordered in the usual way. We also write

$$\xi_i^{(k)} = \lim_{n\to\infty} x_{ni}^{(k)}.$$

Theorem 4.1 now yields directly the following result to which we will have reference in Chapter IV.

THEOREM **4.2**   For every $k \geqq 0$,

$$\xi_i^{(k)} \leq \xi_i^{(k+1)} \leq \xi_{i+1}^{(k)}$$

hence

$$\lim_{i\to\infty} \xi_i^{(k)} = \lim_{i\to\infty} \xi_i = \sigma.$$

Proof From Theorem 4.1 we obtain

$$x_{n+1,i}^{(k)} < x_{ni}^{(k+1)} < x_{n+1,i+1}^{(k)}$$

from which the desired conclusions follow.   ■

We next note an important formula for the partial fraction decomposition of the convergents of the J-fraction (4.1) in the positive-definite case.

THEOREM **4.3**   When the c_n are real and $\lambda_n > 0$ $(n \geqq 1)$, we have

$$\frac{\lambda_1 P_{n-1}^{(1)}(x)}{P_n(x)} = \sum_{k=1}^{n} \frac{A_{nk}}{x - x_{nk}} = \int_{-\infty}^{\infty} \frac{d\psi_n(t)}{x - t}.$$

where A_{nk} is the coefficient in the Gauss quadrature formula corresponding to the zero x_{nk} and where ψ_n is the corresponding distribution function II-(3.2) with jump A_{nk} at the point x_{nk}.

Proof   The coefficients in the partial fraction decomposition

$$\frac{\lambda_1 P_{n-1}^{(1)}(x)}{P_n(x)} = \sum_{k=1}^{n} \frac{a_{nk}}{x - x_{nk}}$$

are given by

$$a_{nk} = \lambda_1 \lim_{x \to x_{nk}} \frac{(x - x_{nk}) P_{n-1}^{(1)}(x)}{P_n(x)} = \frac{\lambda_1 P_{n-1}^{(1)}(x_{nk})}{P_n'(x_{nk})}.$$

Using (4.4), we see that

$$\lambda_1 P_{n-1}^{(1)}(x_{nk}) = -\frac{\lambda_1 \lambda_2 \cdots \lambda_{n+1}}{P_{n+1}(x_{nk})}$$

hence

$$a_{nk} = -\frac{\lambda_1 \lambda_2 \cdots \lambda_{n+1}}{P_{n+1}(x_{nk}) P_n'(x_{nk})}.$$

It is now easy to verify using the Christoffel-Darboux identity that $a_{nk} = A_{nk}$ (I-Ex. 6.2).   ■

Another important representation for $P_n^{(1)}(x)$ is

$$P_n^{(1)}(y) = \frac{1}{\mu_0} \mathscr{L} \left\{ \frac{P_{n+1}(y) - P_{n+1}(x)}{y - x} \right\} \tag{4.6}$$

where it is understood $\mathscr{L}$ operates on x. We leave it to the reader to verify (4.6) by showing that the expression on the right side of the equal sign satisfies the recurrence (4.3).

REMARK It is obvious that $\{P_n^{(1)}(x)\}$ is uniquely determined by $\{P_n(x)\}$. It is also clear that $\{P_n^{(1)}(x)\}$ does not uniquely determine $\{P_n(x)\}$ since $P_n^{(1)}(x)$ is independent of both c_1 and λ_2 (and of course λ_1).

Thus for example the monic Tchebichef polynomials $\hat{U}_n(x) = 2^{-n} U_n(x)$ are the numerator polynomials for both $\{\hat{T}_n(x)\} = \{2^{1-n} T_n(x)\}$ and $\{\hat{U}_n(x)\}$ itself. More generally, $\{\hat{U}_n(x)\}$ is the monic numerator OPS corresponding to $\{V_n(x)\} = \{V_n(c,\lambda; x)\}$ defined by

$$V_n(x) = x V_{n-1}(x) - \frac{1}{4} V_{n-2}(x), \qquad n \geq 3$$

$$V_2(x) = x V_1(x) - (\lambda + \frac{1}{4}) V_0(x)$$

$$V_0(x) = 1, \qquad V_1(x) = x - c.$$

It is not difficult to verify that

$$V_n(c,\lambda; x) = \hat{U}_n(x) - c \hat{U}_{n-1}(x) - \lambda \hat{U}_{n-2}(x)$$

where $\hat{U}_{-1}(x) = \hat{U}_{-2}(x) = 0$.

The preceding discussion barely hints at the connection between J-fractions and orthogonal polynomials. While a general treatment of J-fractions and their relations with moment problems and orthogonal polynomials is beyond the scope of this book, some indication of these relations can be given.

Suppose that $\mathcal{L}$ is positive-definite and that its true interval of orthogonality is bounded. For the corresponding orthogonal polynomials, we can write according to Theorem 4.3

$$\frac{\lambda_1 P_{n-1}^{(1)}(z)}{P_n(z)} = \int_{\xi_1}^{\eta_1} \frac{d\psi_n(x)}{z - x}. \tag{4.7}$$

But according to II-Theorem 3.1 there is a subsequence $\{\psi_{n_k}\}$ which converges to a representative ψ of $\mathcal{L}$. From Helly's second theorem we can then conclude that

$$\lim_{k \to \infty} \frac{\lambda_1 P_{n_k-1}^{(1)}(z)}{P_{n_k}(z)} = \int_{\xi_1}^{\eta_1} \frac{d\psi(x)}{z - x} \qquad \text{for } z \notin [\xi_1, \eta_1].$$

(I-Theorem 2.3 was actually proved for real integrands but there is no difficulty in proving it for complex-valued integrands.)

It follows that if $[\xi_1, \eta_1]$ is bounded, there is a subsequence of the convergents of the J-fraction (4.1) which converges to

$$F(z) = \int_{\xi_1}^{\eta_1} \frac{d\psi(x)}{z - x}, \qquad z \notin [\xi_1, \eta_1], \qquad (4.8)$$

where ψ is the substantially unique representative of $\mathcal{L}$. It was first proved by A. Markov in 1896 that in fact the J-fraction converges uniformly to $F(z)$ on every compact subset of the complex plane that does not intersect the interval $[\xi_1, \eta_1]$.

Now if $F(z)$ is known, then the distribution function ψ can be recovered from $F(z)$ by means of the *Stieltjes inversion formula*:

$$\psi(t) - \psi(s) = -\frac{1}{\pi} \lim_{y \to 0+} \int_s^t \text{Im}\{F(x + iy)\} \, dx. \qquad (4.9)$$

As given by (4.9), ψ is "normalized" by being redefined, if necessary, at points of discontinuity so that

$$\psi(x) = \frac{\psi(x + 0) - \psi(x - 0)}{2}.$$

In the event $[\xi_1, \eta_1]$ is unbounded, the situation is considerably more complex and convergence questions are closely tied to the determinacy problem of the corresponding moment problem. However there is always an analytic function of the form (4.8) associated with a J-fraction and (4.9) remains valid.

Exercises

4.1 Verify (4.6).

4.2 Let $\{P_n(c, x)\}$ be defined by

$$P_n(c, x) = (x - c_n) P_{n-1}(c, x) - \lambda_n P_{n-2}(c, x), \qquad n \geq 2.$$

$$P_0(c, x) = 1, \qquad P_1(c, x) = P_1(x) - c = x - (c_1 + c).$$

Show that $P_n(c, x) = P_n(x) - c P_{n-1}^{(1)}(x)$ (where $P_n(x) = P_n(0, x)$). Assuming the positive-definite case, obtain a separation theorem involving the zeros of $P_n(c, x)$, $P_n(x)$, and $P_{n-1}^{(1)}(x)$

4.3 Let $\{P_n^{\#}(\lambda, x)\}$ be defined by

$$P_n^{\#}(\lambda, x) = (x - c_n) P_{n-1}^{\#}(\lambda, x) - \lambda_n P_{n-2}^{\#}(\lambda, x). \qquad n \geq 3.$$

$$P_2^{\#}(\lambda, x) = (x - c_1) P_1^{\#}(\lambda, x) - (\lambda_2 + \lambda) P_0^{\#}(\lambda, x).$$

$$P_0^{\#}(\lambda, x) = 1. \qquad P_1^{\#}(\lambda, x) = P_1(x).$$

Assuming the positive-definite case, let $\lambda > -\lambda_2$ and obtain a sort of separation theorem involving the zeros of $P_n^{\#}(\lambda, x)$ and $P_n(x)$.

4.4 Let $S_m(x) = xS_{m-1}(x) - \gamma_m S_{m-2}(x)$ where $\gamma_{2m} = a > 0$ and $\gamma_{2m+1} = b > 0$ $(m \geq 1)$.

  (a) Derive the representations

$$S_{2m}(x) = (ab)^{m/2}\left[U_m(z) + \sqrt{\frac{b}{a}}\, U_{m-1}(z)\right]$$

$$S_{2m+1}(x) = (ab)^{m/2} x U_m(z)$$

where $z = (x^2 - a - b)(4ab)^{-1/2}$.

  (b) Conclude from (a) that there is an $h_0 > 0$ such that for $m \neq n$,

$$\mathcal{S}[S_m(x)S_n(x)] = h_0 S_m(0)S_n(0)$$
$$+ \int_E S_m(x)S_n(x)\frac{1}{|x|}\left\{1 - \left(\frac{x^2 - a - b}{4ab}\right)^2\right\}^{1/2} dx = 0,$$

where $E = (-\sqrt{a} - \sqrt{b}, -|\sqrt{a} - \sqrt{b}|) \cup (|\sqrt{a} - \sqrt{b}|, \sqrt{a} + \sqrt{b})$. [The jump h_0 can be determined explicitly: see VI-§13(C).]

4.5 Prove that $c_1 - \xi_1 \geq \sum_{j=1}^{\infty} (\xi_{j+1} - \xi_j^{(1)}) \geq 0$.

4.6 A continued fraction of the form

$$\frac{\gamma_1|}{|x} - \frac{\gamma_2|}{|1} - \frac{\gamma_3|}{|x} - \frac{\gamma_4|}{|1} - \cdots, \qquad \gamma_i > 0,$$

is called an S-fraction (after Stieltjes, who based his study of the moment problem on continued fractions having a similar form). Let $B_0 = 1$ and $A_n(x)/B_n(x)$ denote the nth convergent with $B_n(x)$ monic. Prove that $\{B_{2n}(x)\}_{n=0}^{\infty}$ is an OPS whose zeros lie in $[0, \infty)$ and that $\{x^{-1}B_{2n+1}(x)\}_{n=0}^{\infty}$ is the corresponding monic KPS with K-parameter 0.

5 Chain Sequences

We undertake a formal study of real sequences having the special form of the numerators of the continued fractions that appeared in the lemmas and theorem of §3. The appearance of such a sequence in I-Theorem 9.2 suggests the role such sequences will play in our study of orthogonal polynomials in Chapter IV.

DEFINITION 5.1 A sequence $\{a_n\}_{n=1}^{\infty}$ is called a *chain sequence* if there exists a sequence $\{g_k\}_{k=0}^{\infty}$ such that

  (i) $0 \leq g_0 < 1, \qquad 0 < g_n < 1, \qquad n \geq 1$

  (ii) $a_n = (1 - g_{n-1})g_n, \qquad\qquad n = 1, 2, 3, \ldots$ (5.1)

$\{g_k\}$ is called a *parameter sequence* for $\{a_n\}$. g_0 is called an *initial parameter*.

The systematic development of the theory of chain sequences is due to H. S. Wall. The definition we have adopted above is more restrictive than that of Wall who only requires in place of (i),

$$(i')   0 \leq g_n \leq 1   (n \geq 0).$$

To be consistent with Wall's definition then, we should perhaps speak of *positive* chain sequences when (i) is satisfied. However, in our applications of this concept, we will be concerned with positive chain sequences only and hence will omit the adjective. In most cases, simple modifications in the statements and proofs of theorems will provide corresponding results for the more general case. These can be found in Wall's book [2].

While it is trivial to construct examples of chain sequences, it is frequently difficult to determine if a given sequence is a chain sequence. In some cases, one can determine a parameter sequence inductively by taking $g_0 = 0$, $g_1 = a_1$, and attempting to solve (5.1) successively for the g_i. For example, if one tries this with the constant sequence, $\{1/4\}$, one finds

$$g_0 = 0, g_1 = \frac{1}{4}, g_2 = \frac{1}{3}, g_3 = \frac{3}{8}, \ldots$$

from which it is easily deduced that $g_n = n/[2(n + 1)]$. Thus $\{1/4\}$ is a chain sequence. However this chain sequence also has the much simpler parameter sequence $\{1/2\}$.

More generally, we note the identities

$$a = \left(1 - \frac{1 - \sqrt{1 - 4a}}{2}\right)\frac{1 - \sqrt{1 - 4a}}{2}$$

$$= \left(1 - \frac{1 + \sqrt{1 - 4a}}{2}\right)\frac{1 + \sqrt{1 - 4a}}{2}.$$

Thus the constant sequence $\{a\}$ is a chain sequence if $0 < a \leq 1/4$.

We note from the above examples that a chain sequence need not have a unique parameter sequence.

THEOREM 5.1   Let $\{a_n\}$ be a chain sequence and let $\{g_k\}$ and $\{h_k\}$ both be parameter sequences for $\{a_n\}$. Then

$$g_k < h_k   \text{for } k \geq 1 \text{ if and only if } g_0 < h_0.$$

Proof If $(1 - g_{n-1})g_n = a_n = (1 - h_{n-1})h_n$, then

$$\frac{g_k}{h_k} = \frac{1 - h_{k-1}}{1 - g_{k-1}}.$$

Thus $0 < g_k < h_k < 1$ if and only if $0 < g_{k-1} < h_{k-1} < 1$ from which the theorem follows. ∎

THEOREM 5.2 Let $\{a_n\}$ be a chain sequence. If $\{a_n\}$ has a parameter sequence $\{g_k\}$ such that $g_0 > 0$, then for each h_0 such that $0 \leq h_0 < g_0$, there is a corresponding parameter sequence $\{h_n\}$.

Proof If $0 \leq h_0 < g_0$, then

$$0 < \frac{a_1}{1 - h_0} < \frac{a_1}{1 - g_0} = g_1.$$

Thus $a_1/(1 - h_0) = h_1 < g_1$ so that

$$a_1 = (1 - h_0)h_1, \qquad 0 < h_1 < g_1.$$

Similarly, if $0 < h_k < g_k$, then

$$\frac{a_{k+1}}{1 - h_k} < \frac{a_{k+1}}{1 - g_k} = g_{k+1}$$

so there is a unique h_{k+1} such that

$$a_{k+1} = (1 - h_k)h_{k+1}, \qquad 0 < h_{k+1} < g_{k+1}.$$

By induction it follows that a unique parameter sequence $\{h_k\}$ exists. ∎

According to Theorem 5.2, every chain sequence has a parameter sequence $\{m_k\}$ such that $m_0 = 0$ and by Theorem 5.1 we then have $m_n < g_n$ for every other parameter sequence $\{g_k\}$.

DEFINITION 5.2 Let $\{a_n\}$ be a chain sequence. A parameter sequence $\{m_k\}$ is called its *minimal parameter sequence* if $m_0 = 0$.

If the minimal parameter sequence is the only parameter sequence for $\{a_n\}$, then $\{a_n\}$ is said to *determine its parameters uniquely*. If $\{a_n\}$ does not determine its parameters uniquely, then Theorem 5.2 shows that the initial parameters form a connected set.

DEFINITION 5.3   Let $\{a_n\}$ be a chain sequence. A parameter sequence $\{M_k\}$ is called its *maximal parameter sequence* if $M_k > g_k$ $(k \geqq 0)$ for every other parameter sequence $\{g_k\}$.

THEOREM 5.3   Every chain sequence has a maximal parameter sequence.

Proof   If $\{a_n\}$ is a chain sequence that determines its parameters uniquely, then its minimal parameter sequence is clearly its maximal parameter sequence. Otherwise, let M_0 denote the least upper bound of the set of all initial parameters for $\{a_n\}$. Then $0 < M_0 \leqq 1$.

Now for each x such that $0 \leqq x < M_0$, there is a parameter sequence $\{g_n(x)\}$ such that $g_0(x) = x$. According to Theorem 5.1, for each fixed value of n, g_n is an increasing function. Thus we can define

$$M_n = \lim_{x \to M_0^-} g_n(x).$$

We clearly have $0 < M_n \leqq 1$ and also

$$0 < (1 - M_{n-1})M_n = a_n < 1, \qquad n \geqq 1.$$

Thus $0 < M_n < 1$ for every n and $\{M_k\}$ is a parameter sequence which must be maximal because of the definition of M_0. ∎

We will later obtain an explicit representation of the maximal parameters by means of continued fractions as well as an important criterion for determining whether or not a given parameter sequence is maximal. For now we content ourselves with the above existence theorems and results that can be obtained without recourse to continued fractions.

Throughout the remainder of this chapter, $\{a_n\}$ will denote a chain sequence and $\{m_k\}$ and $\{M_k\}$, respectively, will denote its minimal and maximal parameter sequences. We will also use the notation

$$b_n^{(k)} = b_{n+k}$$

where $\{b_n\}$ is any sequence. In the notation, $\{b_n^{(k)}\}$, it will always be the case that the sequence is indexed by the *subscript* but this will be indicated explicitly if there is any danger of confusion.

THEOREM 5.4   (i) $\{a_n^{(1)}\}$ is a chain sequence with parameter sequence $\{g_k^{(1)}\}_{k=0}^{\infty}$ where $\{g_k\}_{k=0}^{\infty}$ is any parameter sequence for $\{a_n\}$.

(ii) If $\{m_{1k}\}_{k=0}^{\infty}$ denotes the minimal parameter sequence for $\{a_n^{(1)}\}$, then $m_{1k} < m_k^{(1)}$ $(k \geqq 0)$.

(iii) $\{M_k^{(1)}\}_{k=0}^{\infty}$ is the maximal parameter sequence for $\{a_n^{(1)}\}$.

Proof Since

$$a_n^{(1)} = a_{n+1} = (1 - g_n)g_{n+1} = (1 - g_{n-1}^{(1)})g_n^{(1)}, \qquad n \geq 1,$$

(i) follows at once.

Since

$$m_{10} = 0 < m_1 = m_0^{(1)},$$

Theorem 5.1 yields (ii).

To prove (iii), let $\{M_k'\}$ denote the maximal parameter sequence for $\{a_n^{(1)}\}$. Since $\{M_k^{(1)}\}$ is a parameter sequence for $\{a_n^{(1)}\}$, we have $0 < M_0^{(1)} \leq M_0'$, hence

$$(1 - M_0)M_1 \leq (1 - M_0)M_0'.$$

Thus there exists an $M^* \geq M_0$ such that

$$(1 - M^*)M_0' = (1 - M_0)M_1 = a_1.$$

It now follows that $\{M^*, M_0', M_1', M_2', \ldots\}$ is a parameter sequence for $\{a_n\}$ and this is impossible unless $M^* = M_0$. But then we must have $M_k' = M_{k+1}$ ($k \geq 0$) and this says that $\{M_k'\} = \{M_k^{(1)}\}$. ∎

THEOREM 5.5 Let $\{b_n\}$ be a chain sequence with parameter sequence $\{h_k\}$. If

$$a_n \leq b_n, \qquad n \geq 1,$$

then

$$m_k \leq h_k \leq M_k, \qquad k \geq 0.$$

Proof If $a_1 \leq b_1 = (1 - h_0)h_1$, then there is a g_1 such that

$$a_1 = (1 - h_0)g_1, \qquad 0 < g_1 \leq h_1.$$

Similarly, if $g_0(= h_0), g_1, g_2, \ldots, g_n$ exist such that

$$a_k = (1 - g_{k-1})g_k, \qquad 0 < g_k \leq h_k, \qquad k = 1, 2, \ldots, n,$$

then

$$a_{n+1} \leq b_{n+1} = (1 - h_n)h_{n+1} \leq (1 - g_n)h_{n+1}.$$

It then follows that there is a g_{n+1} such that

$$a_{n+1} = (1 - g_n)g_{n+1}, \qquad 0 < g_{n+1} \leq h_{n+1}.$$

This proves by induction that $\{a_n\}$ has a parameter sequence $\{g_k\}$ such that

$$g_0 = h_0, \qquad 0 < g_n \leq h_n \qquad \text{for } n \geq 1.$$

This shows that $m_n \leq h_n$ $(n \geq 0)$ and also that $g_0 = h_0 \leq M_0$. Applying this reasoning to $\{a_n^{(k)}\}$ and $\{b_n^{(k)}\}$, we conclude that

$$h_k = h_0^{(k)} \leq M_0^{(k)} = M_k, \qquad k \geq 1. \qquad \blacksquare$$

COROLLARY   If for some $N \geq 1$, we have in addition

$$a_N < b_N,$$

then

$$m_k < h_k \qquad (k \geq N) \quad \text{and} \quad h_k < M_k \qquad (0 \leq k < N).$$

In particular, if $\{a_n\}$ is dominated by a second chain sequence, then $\{a_n\}$ cannot determine its parameters uniquely.

The proof is left as an exercise.

THEOREM 5.6   Let $\{a_n\}$ be non-decreasing. Then $\{m_k\}$ is strictly increasing and $\{M_k\}$ is non-increasing.

Proof   Let $\{m_{1k}\}_{k=0}^{\infty}$ denote the minimal parameter sequence for $\{a_n^{(1)}\}$. Now by Theorem 5.4, $m_{1k} < m_k^{(1)}$, and $\{M_k^{(1)}\}$ is the maximal parameter sequence for $\{a_n^{(1)}\}$.
Now if $a_n \leq a_n^{(1)}$, then according to Theorem 5.5,

$$m_k \leq m_{1k}, \qquad M_k^{(1)} \leq M_k.$$

Thus $m_k < m_{k+1}$ and $M_{k+1} \leq M_k$ $(k \geq 0)$. $\qquad \blacksquare$

COROLLARY 1   If $a_n = 1/4$ $(n \geq 1)$, then $M_k = 1/2$ $(k \geq 0)$.

Proof By Theorem 5.6, $M_{k+1} \leqq M_k$ so

$$\frac{1}{4} = (1 - M_k)M_{k+1} \leqq (1 - M_k)M_k \leqq \frac{1}{4},$$

Thus $(1 - M_k)M_k = 1/4$. ∎

COROLLARY 2   If $a_n \geqq 1/4$ for $n \geqq N$, then $\lim_{n\to\infty} a_n = 1/4$. Hence if $b_n \geqq b > 1/4$ for $n \geqq N$, then $\{b_n\}$ is not a chain sequence.

Proof In view of Theorem 5.4, it is sufficient to consider $N = 1$. We have noted that the minimal parameters for $\{1/4\}$ are

$$m_k = \frac{k}{2(k + 1)}, \qquad k \geqq 0.$$

Therefore, if $\{g_k\}$ is a parameter sequence for $\{a_n\}$ and $a_n \geqq 1/4$, then by Theorem 5.5 and Corollary 1 of Theorem 5.6, $g_k \to 1/2$ as $k \to \infty$. ∎

We next prove Wall's useful "comparison test" for chain sequences.

THEOREM 5.7   (Comparison test) If $0 < c_n \leqq a_n$ for $n \geqq 1$, then $\{c_n\}$ is also a chain sequence.

Proof Define $h_0 = 0$ and $h_1 = c_1$ so that $h_1 \leqq a_1 = m_1$. Then

$$c_1 = (1 - h_0)h_1, \qquad 0 < h_1 \leqq m_1.$$

Now if

$$c_k = (1 - h_{k-1})h_k, \qquad 0 < h_k \leqq m_k, \qquad k = 1, 2, \ldots, n,$$

then

$$c_{n+1} \leqq a_{n+1} = (1 - m_n)m_{n+1} \leqq (1 - h_n)m_{n+1}.$$

Thus there exists an h_{n+1} such that $0 < h_{n+1} \leqq m_{n+1}$ and $c_{n+1} = (1 - h_n)h_{n+1}$. By induction, it follows that $\{c_n\}$ is a chain sequence. ∎

We see from the preceding that $\{c_n\}$ is a chain sequence if $0 < c_n \leqq 1/4$ ($n \geqq 1$). Slightly more generally, $\{c_n\}$ is a chain sequence if $0 < c_1 \leqq 1/2$ and $0 < c_n \leqq 1/4$ for $n \geqq 2$ (Ex. 5.1). It then follows that a geometric sequence $\{r^n\}_{n=1}^{\infty}$ is a chain sequence if $0 < r \leqq 1/2$.

If $\{a_n\}$ determines its parameters uniquely, no other chain sequence can dominate $\{a_n\}$. However, if $\{a_n\}$ does not determine its parameters uniquely, then it is easy to construct a new chain sequence $\{b_n\}$ that dominates $\{a_n\}$. For example, we can take $b_n = (1 - h_{n-1})h_n$ where either

$$\text{(a)} \qquad h_k = \sqrt{m_k M_k}$$

or

$$\text{(b)} \qquad h_k = \lambda m_k + (1 - \lambda)M_k, \qquad 0 < \lambda < 1.$$

In case (a), we have

$$\begin{aligned} b_n^2 &= (1 - 2\sqrt{m_{n-1} M_{n-1}} + m_{n-1} M_{n-1})m_n M_n \\ &> (1 - m_{n-1} - M_{n-1} + m_{n-1} M_{n-1})m_n M_n \\ &= (1 - m_{n-1})(1 - M_{n-1})m_n M_n = a_n^2. \end{aligned}$$

Case (b) is left to the exercises. However we note that if we take $a_n = 1/4$ and $\lambda = 1/2$ in (b), we get the chain sequence $\{b_n\}$ where

$$b_n = \frac{1}{4} + \frac{1}{16n(n + 1)}. \tag{5.2}$$

Note that in the last example,

$$\sum_{n=1}^{\infty} \left(b_n - \frac{1}{4}\right) = \frac{1}{16}.$$

More generally, we have the following necessary condition for chain sequences.

THEOREM 5.8

$$\sum_{k=1}^{n} \left(\sqrt{a_k} - \frac{1}{2}\right) < \frac{m_n}{2}.$$

In particular, if $a_n \geqq 1/4$ ($n \geqq 1$), then

$$0 \leqq \sum_{k=1}^{n} \left(a_k - \frac{1}{4}\right) < \frac{3}{8}. \tag{5.3}$$

Proof   In general we have

$$\sqrt{a_n} = \sqrt{(1 - m_{n-1})m_n} \leqq \frac{(1 - m_{n-1}) + m_n}{2},$$

hence

$$2 \sum_{k=1}^{n} \sqrt{a_k} \leqq n + m_n - m_0 = n + m_n.$$

Therefore

$$\sum_{k=1}^{n} \left(\sqrt{a_k} - \frac{1}{2} \right) \leqq \frac{m_n}{2}. \tag{5.4}$$

In particular, if $a_n \geqq 1/4$ for $n \geqq 1$, then by Theorem 5.5,

$$\frac{n}{2(n+1)} \leqq m_n \leqq \frac{1}{2}.$$

Also

$$0 \leqq a_k - \frac{1}{4} = \left(\sqrt{a_k} - \frac{1}{2} \right)\left(\sqrt{a_k} + \frac{1}{2} \right) \leqslant \frac{3}{2}\left(\sqrt{a_k} - \frac{1}{2} \right)$$

with equality holding only if $a_k = 1/4$, so (5.3) follows from (5.4). ∎

COROLLARY If $b_n \geqq 1/4$ for $n \geqq N$ and

$$\sum_{n=N}^{\infty} \left(b_n - \frac{1}{4} \right) = \infty$$

then $\{b_n^{(k)}\}_{n=1}^{\infty}$ is not a chain sequence for any k.

Proof By Theorem 5.8, $\{b_n^{(k)}\}$ is not a chain sequence for $k \geqq N$. By Theorem 5.4 then, it is not a chain sequence for any k. ∎

Exercises

5.1 Show that $\{1/2, 1/4, 1/4, 1/4, \ldots\}$ is a chain sequence that determines its parameters uniquely.

5.2 Show that $\{(n+1)^{-1}\}_{n=1}^{\infty}$ is a chain sequence by finding its minimal parameter sequence.

5.3 Let $0 < a \leqq 1/4$. Prove that the minimal parameters of $\{a\}$ are given by $m_n = \frac{1}{2}[1 - \sqrt{1 - 4a}](1 - S_n^{-1})$ where

$$S_n = \sum_{k=0}^{n} \left(\frac{1 + \sqrt{1 - 4a}}{1 - \sqrt{1 - 4a}} \right)^k \tag{Wall [2]}.$$

5.4 Let $a_n \geq 1/4$. If $\{g_k\}_{k=0}^{\infty}$ is a parameter sequence for $\{a_n\}$ and $g_N = 1/2$, prove that $g_k = 1/2$ for all $k > N$.

5.5 Prove that if $a_n \geq 1/4$ for every n, then every parameter sequence is non-decreasing. (Wall [2])

5.6 Show that the constant $3/8$ in (5.3) can be replaced by $(1 + \sqrt{2})/8$. Conjecture: the constant can be replaced by $1/4$ if $<$ is replaced by $\leq$.

5.7 Prove that if $b_n = 1/4 + \epsilon_n$ where $\epsilon_n \geq 0$ and $\sum_{k=1}^{\infty} k\epsilon_k \leq 1/4$, then $\{b_n\}$ is a chain sequence. This result and Ex. 5.6 suggest the question: by how much can the terms of a chain sequence exceed $1/4$?

5.8 Prove that $b_n > a_n$ if $b_n = (1 - h_{n-1})h_n$ where $h_n = \lambda m_n + (1 - \lambda)M_n$, $0 < \lambda < 1$.

5.9 Let $\{c_n\}$ be a sequence of positive numbers such that $\sum_{k=1}^{n} c_k < 1$ for $n \geq 1$. Prove that $\{c_n\}$ is a chain sequence. (Wall [2]).

5.10 (a) Prove that $\{a_{n-1}\}_{n=1}^{\infty}$ is also a chain sequence if and only if $M_0 > 0$ and $0 < a_0 \leq M_0$.
     (b) If $\{a_{n-1}\}_{n=1}^{\infty}$ is a chain sequence, show that its maximal parameter sequence is $\{M_{k-1}\}_{k=0}^{\infty}$ where $M_{-1} = 1 - a_0/M_0$.

5.11 Show that M_n is independent of $a_1, \ldots, a_n$ whereas m_n is independent of $a_{n+1}, a_{n+2}, a_{n+3}, \ldots$.

5.12 Let $\{a_{p,n}\}_{n=1}^{\infty}$ be a chain sequence for each positive integer p. Suppose that $\lim_{p \to \infty} a_{p,n} = a_n > 0$ $(n = 1, 2, 3, \ldots)$. Prove that $\{a_n\}$ is also a chain sequence. (Haddad [1])

5.13 Prove that $\{b_n\}$ is a chain sequence if and only if $b_n > 0$ and

$$\sum_{i=1}^{n+1} x_i^2 - 2 \sum_{i=1}^{n} \sqrt{b_i}\, x_i x_{i+1} \geq 0, \qquad n = 1, 2, 3, \ldots$$

for all real numbers x_i. (Wall [2]).

6 Additional Results on Chain Sequences

We now obtain Wall's important characterization of the maximal parameters of a chain sequence. We maintain the conventions of the preceding section so that $\{a_n\}$ is a chain sequence, $\{m_k\}$ and $\{M_k\}$ are its minimal and maximal parameter sequences, respectively. We also will denote by $\{m_{kn}\}_{n=0}^{\infty}$, the minimal parameter sequence of the chain sequence $\{a_n^{(k)}\}_{n=1}^{\infty}$.

According to Lemma 3.2, we have

$$P_k = 1 - \frac{a_1^{(k)}|}{|\,1} - \frac{a_2^{(k)}|}{|\,1} - \frac{a_3^{(k)}|}{|\,1} - \cdots, \qquad k = 0, 1, 2, \ldots, \qquad (6.1)$$

where

$$P_k = \frac{1}{1 + L_k}, \qquad L_k = \sum_{n=1}^{\infty} \frac{m_{k1} m_{k2} \cdots m_{kn}}{(1 - m_{k1})(1 - m_{k2}) \cdots (1 - m_{kn})}.$$

It is clear that $0 \leqq P_k < 1$. If we consider (6.1) for two consecutive values of k, we find

$$P_k = 1 - \frac{a_{k+1}}{P_{k+1}}.$$

It follows that $P_{k+1} \neq 0$ so that

$$(1 - P_k)P_{k+1} = a_{k+1}, \qquad 0 \leqq P_0 < 1, \qquad 0 < P_{k+1} < 1 \qquad \text{for } k \geqq 0.$$

Thus $\{P_k\}$ is a parameter sequence for $\{a_n\}$.

THEOREM 6.1 $\{P_k\}$ defined by (6.1) is the maximal parameter sequence for $\{a_n\}$.

Proof If $\{g_k\}$ is any parameter sequence for $\{a_n\}$, then by Theorem 3.1,

$$P_0 = g_0 + (1 - g_0)(1 + G)^{-1},$$

$$G = \sum_{n=1}^{\infty} \frac{g_1 g_2 \cdots g_n}{(1 - g_1)(1 - g_2) \cdots (1 - g_n)}. \tag{6.2}$$

Thus $P_0 \geqq g_0$ so by Theorem 5.1, $\{P_k\}$ is the maximal parameter sequence. ■

THEOREM 6.2 A parameter sequence $\{g_k\}$ is the maximal parameter sequence for $\{a_n\}$ if and only if

$$\sum_{n=1}^{\infty} \frac{g_1 g_2 \cdots g_n}{(1 - g_1)(1 - g_2) \cdots (1 - g_n)} = \infty.$$

Proof From (6.2) we have $P_0 = g_0$ if and only if $G = \infty$. ■

We had noted earlier that if $a_n = a$, $0 < a \leqq 1/4$, then $\{a_n\}$ has parameter sequences $\{g_k\}$ and $\{h_k\}$ where

$$g_k = \frac{1}{2}[1 - \sqrt{1 - 4a}], \qquad h_k = \frac{1}{2}[1 + \sqrt{1 - 4a}]. \tag{6.3}$$

Theorem 6.2 shows that $\{h_k\}$ is actually the maximal parameter sequence. We will need to know later $\lim_{k \to \infty} m_k$. Although $\{m_k\}$ can be found explicitly (see Ex. 5.3), the next theorem circumvents these computations.

THEOREM **6.3**   If $\{g_n\}$ is any non-maximal parameter sequence for $\{a_n\}$, then

$$\lim_{n \to \infty} \frac{m_n}{g_n} = 1.$$

Proof   Let $d_n = g_n - m_n$. From $a_n = (1 - m_{n-1})m_n = (1 - g_{n-1})g_n$, follows

$$d_n = g_n d_{n-1} + m_{n-1} d_n.$$

Thus

$$0 \leq d_n = \frac{g_n d_{n-1}}{1 - m_{n-1}}$$

$$= \prod_{k=1}^{n} \frac{g_k}{1 - m_{k-1}} d_0 = g_n \prod_{k=0}^{n-1} \frac{g_k}{1 - m_k},$$

$$0 \leq 1 - \frac{m_n}{g_n} = \prod_{k=0}^{n-1} \frac{g_k}{1 - m_k} \leq \prod_{k=0}^{n-1} \frac{g_k}{1 - g_k}.$$

Referring to Theorem 6.2 we see that if $\{g_k\}$ is not the maximal parameter sequence, then

$$\lim_{n \to \infty} \prod_{k=0}^{n} \frac{g_k}{1 - g_k} = 0$$

which yields the desired conclusion.   ∎

COROLLARY   If $a_n = a,\ 0 < a \leq 1/4$, then

$$\lim_{n \to \infty} m_n = \frac{1}{2}[1 - \sqrt{1 - 4a}], \qquad M_n = \frac{1}{2}[1 + \sqrt{1 - 4a}]. \qquad (6.4)$$

The corollary shows that the conclusion of Theorem 6.3 need not hold if g_n is replaced by M_n.

If $\{g_n\}$ is a parameter sequence for $\{a_n\}$ and $g_n \to g$, then $a_n \to (1 - g)g \leq 1/4\ (n \to \infty)$. We next show that the converse is true.

THEOREM **6.4**   Let

$$\lim_{n \to \infty} a_n = a.$$

Then $0 \leqq a \leqq 1/4$ and

$$\lim_{n \to \infty} M_n = \frac{1}{2}[1 + \sqrt{1 - 4a}]. \qquad (6.5)$$

Moreover, if $M_0 > 0$ (that is, $m_n \neq M_n$), then

$$\lim_{n \to \infty} m_n = \frac{1}{2}[1 - \sqrt{1 - 4a}]. \qquad (6.6)$$

Proof If $a > 1/4$, then there is an N such that $a_n \geqq a' > 1/4$ for $n \geqq N$. According to Corollary 2, Theorem 5.6, $\{a_n\}$ would not be a chain sequence. Thus we must have $0 \leqq a \leqq 1/4$.

Now suppose first that $0 < a < 1/4$ and choose $\epsilon > 0$ such that

$$0 < a - \epsilon < a + \epsilon < 1/4.$$

Then there exists $k = k(\epsilon)$ such that

$$a - \epsilon < a_n^{(k)} < a + \epsilon, \qquad n = 1, 2, 3, \ldots . \qquad (6.7)$$

Let $\{m_n'\}$ and $\{M_n'\}$, $\{m_n''\}$ and $\{M_n''\}$ denote the minimal and maximal parameters, respectively, of the chain sequences $\{a - \epsilon\}$ and $\{a + \epsilon\}$. Then by Theorem 5.5 and (6.7) we have

$$m_n' \leqq m_{kn} \leqq m_n'', \qquad M_n'' \leqq M_n^{(k)} \leqq M_n'. \qquad (6.8)$$

(Recall that $\{m_{kn}\}_{n=0}^{\infty}$ denotes the minimal parameter sequence for $\{a_n^{(k)}\}$.)

From the corollary to Theorem 6.3 we thus obtain

$$\frac{1}{2}[1 + \sqrt{1 - 4(a + \epsilon)}] \leqq M_n^{(k)} \leqq \frac{1}{2}[1 + \sqrt{1 - 4(a - \epsilon)}],$$

$$\frac{1}{2}[1 - \sqrt{1 - 4(a - \epsilon)}] \leqq \liminf_{n \to \infty} m_{kn}$$

$$\limsup_{n \to \infty} m_{kn} \leqq \frac{1}{2}[1 - \sqrt{1 - 4(a + \epsilon)}].$$

Since these are valid for arbitrarily small $\epsilon > 0$, we conclude

$$\lim_{n \to \infty} m_{kn} = \frac{1}{2}[1 - \sqrt{1 - 4a}], \qquad \lim_{n \to \infty} M_n = \frac{1}{2}[1 + \sqrt{1 - 4a}].$$

In particular, this establishes (6.5). If $M_0 > 0$, then $\{m_k\}$ is not the maximal parameter sequence for $\{a_n\}$, hence $\{m_n^{(k)}\}$ is not the maximal parameter sequence for $\{a_n^{(k)}\}$. Thus by Theorem 6.3,

$$\lim_{n\to\infty} m_n^{(k)} = \lim_{n\to\infty} m_{kn} = \frac{1}{2}[1 - \sqrt{1 - 4a}]$$

and this yields (6.6).

For the case $a = 0$, replace $a - \epsilon$ by 0 in (6.7) and use the inequalities

$$0 < m_{kn} \leqq m_n'', \qquad M_n'' \leqq M_n^{(k)} < 1$$

in place of (6.8). The corollary to Theorem 6.3 then gives

$$0 \leqq \limsup_{n\to\infty} m_{kn} \leqq \frac{1}{2}[1 - \sqrt{1 - 4\epsilon}]$$

$$\frac{1}{2}[1 + \sqrt{1 - 4\epsilon}] \leqq \liminf_{n\to\infty} M_n^{(k)} \leqq 1,$$

from which (6.5) and (6.6) follow.

Finally, for the case $a = 1/4$, we need only the inequality

$$\frac{1}{4} - \epsilon < a_n^{(k)}.$$

This and Theorem 5.5 then yields

$$m_n' \leqq m_n^{(k)} \leqq M_n^{(k)} \leqq \frac{1}{2}[1 + \sqrt{4\epsilon}].$$

Since $m_n' \to \frac{1}{2}[1 - \sqrt{4\epsilon}]$, we conclude that

$$\lim_{n\to\infty} m_n = \lim_{n\to\infty} M_n = \frac{1}{2}. \qquad \blacksquare$$

We conclude our discussion of chain sequences with a brief look at periodic sequences.

THEOREM 6.5 If $\{a_n\}$ is periodic with period p, then $\{M_k\}$ is also periodic with period p while $\{m_k\}$ is "almost periodic" in the sense that

$$\lim_{n\to\infty} m_{np+k} = \mu_k, \qquad k = 0, 1, \ldots, p - 1.$$

Proof If $\{a_n^{(p)}\} = \{a_n\}$, then $\{M_k^{(p)}\}$ is the maximal parameter sequence for $\{a_n\}$, hence $\{M_k^{(p)}\} = \{M_k\}$ as asserted.

Further, $\{m_k^{(p)}\}$ is a non-minimal parameter sequence for $\{a_n\}$, hence

$$m_k < m_k^{(p)}, \qquad k = 0, 1, 2, \ldots.$$

Thus $\{m_{np+k}\}_{n=0}^{\infty}$ is increasing and hence convergent.   ■

Theorem 6.5 shows that a necessary as well as obviously sufficient condition for a periodic sequence $\{b_n\}$, with period p, to be a chain sequence is that the system of equations

$$b_k = (1 - x_{k-1})x_k, \qquad k = 1, 2, \ldots, p,$$

has a solution $(x_0, x_1, \ldots, x_p)$ such that

$$x_0 = x_p, \qquad 0 < x_i < 1, \qquad i = 1, 2, \ldots, p.$$

In particular, for $p = 2$ we have

THEOREM **6.6**   Let $\{b_n\}$ be periodic with period 2, $b_n > 0$. Then $\{b_n\}$ is a chain sequence if and only if

$$\sqrt{b_1} + \sqrt{b_2} \leqq 1. \tag{6.9}$$

Moreover, $\mu_i < M_i$ $(i = 1, 2)$ if and only if $\sqrt{b_1} + \sqrt{b_2} < 1$. (Here μ_i is as in Theorem 6.5.)

Proof   The equations

$$b_1 = (1 - x_0)x_1, \qquad b_2 = (1 - x_1)x_0,$$

have the solutions

$$x_0 = \frac{1}{2}[1 - b_1 + b_2 \pm \sqrt{(1 - b_1 + b_2)^2 - 4b_2}]$$

$$x_1 = \frac{1}{2}[1 - b_2 + b_1 \pm \sqrt{(1 - b_2 + b_1)^2 - 4b_1}].$$

Now in general, for $0 \leqq u, v \leqq 1$, the inequality

$$\sqrt{u} + \sqrt{v} \leqq 1 \tag{6.10}$$

is equivalent to

$$(1 - u + v)^2 - 4v \geqq 0 \quad \text{and} \quad (1 - v + u)^2 - 4u \geqq 0,$$

with strict inequality holding in (6.10) if and only if strict inequality holds in both of the following inequalities. Thus x_0 and x_1 are real if and only if (6.9) holds. There are two solutions if and only if strict inequality holds in (6.9).

It can be verified routinely that if they are real, then they satisfy $0 < x_i < 1$ $(i = 1, 2)$.   ■

Exercises

6.1 Show that if $\{g_n\}$ is not the maximal parameter sequence for $\{a_n\}$, then the infinite product $\prod_{k=1}^{\infty} m_k/g_k$ converges absolutely.

6.2 (a) Prove that if

$$\lim_{n \to \infty} \frac{m_n}{M_n} = L < 1,$$

then

$$\lim_{n \to \infty} m_n = \frac{L}{1 + L} \quad \text{and} \quad \lim_{n \to \infty} M_n = \frac{1}{1 + L}.$$

(b) Find $\{m_n\}$ and $\{M_n\}$ if

$$a_1 = \frac{1}{4}, \qquad a_{2n} = \frac{1}{(n + 1)^2}, \qquad a_{2n+1} = \frac{n}{n + 2}, \qquad n = 1, 2, 3, \dots$$

and observe that $\lim_{n \to \infty} m_n/M_n = 1$ although neither $\{m_n\}$ nor $\{M_n\}$ converge.

6.3 With the notation of Theorem 6.5, show that $\{\mu_k\}$ is a parameter sequence for $\{a_n\}$.

6.4 (a) Prove that if $0 < b_i < 1$, the system of equations

$$b_i = (1 - x_{i-1})x_i, \qquad i = 1, 2, \dots, p$$

has at most two solutions satisfying the conditions

$$x_0 = x_p, \qquad 0 < x_i < 1, \qquad i = 1, 2, \dots, p. \tag{*}$$

(b) Prove that a necessary condition for the above system to have a solution satisfying (*) is

$$\sqrt{b_1} + \sqrt{b_2} + \cdots + \sqrt{b_p} \leq \frac{p}{2}$$

6.5 Show that $\{a_n\}$ is a chain sequence if $\sqrt{a_k} + \sqrt{a_{k+1}} \leq 1$ $(k \geq 1)$. (The example in (5.2) shows this condition is not necessary.) (Haddad [3])

CHAPTER IV

The Recurrence Formula and Properties of Orthogonal Polynomials

1 Introduction

We return to the study of orthogonal polynomials with our primary objective the investigation of the properties of an OPS as determined by the behavior of the coefficients in the classical three term recurrence formula I-(4.6).

We recall that according to I-Theorems 4.1 and 4.4, a necessary and sufficient condition for a sequence of monic polynomials, $\{P_n(x)\}$, to be an OPS with respect to a positive-definite moment functional $\mathcal{L}$ is that there are constants c_n and λ_n such that

$$P_n(x) = (x - c_n)P_{n-1}(x) - \lambda_n P_{n-2}(x), \qquad n = 1, 2, 3, \ldots,$$

$$P_{-1}(x) = 0, \qquad P_0(x) = 1, \tag{1.1}$$

$$c_n \text{ real}, \qquad \lambda_n > 0 \quad (n \geq 1), \qquad \mathcal{L}[1] = \mu_0 = \lambda_1.$$

Throughout this chapter, $\{P_n(x)\}$ will denote the OPS satisfying (1.1) and $\mathcal{L}$ will denote the positive-definite moment functional that is uniquely determined by

$$\mathcal{L}[P_m(x)P_n(x)] = \lambda_1\lambda_2\cdots\lambda_{n+1}\delta_{mn}, \qquad m, n = 0, 1, 2, \ldots \tag{1.2}$$

As before, we will denote the zeros of $P_n(x)$ by x_{ni} with

$$x_{n1} < x_{n2} < \cdots < x_{nn}.$$

We also maintain the notation

$$\xi_i = \lim_{n\to\infty} x_{ni}, \qquad \eta_j = \lim_{n\to\infty} x_{n,n-j+1}$$

$$\sigma = \lim_{i\to\infty} \xi_i, \qquad \tau = \lim_{j\to\infty} \eta_j. \tag{1.3}$$

Exercise

1.1 Let $xp_n(x) = b_n p_{n-1}(x) + a_n p_n(x) + b_{n+1} p_{n+1}(x)$, where $p_{-1}(x) = 0$, $p_0(x) = b_0^{-1}$, a_n is real and $b_n > 0$ $(n \geq 0)$. Show that $\{p_n(x)\}$ is an orthonormal polynomial sequence with respect to some positive-definite moment functional.

2   Chain Sequences and Orthogonal Polynomials

The theory of chain sequences developed in Chapter III can be used to obtain a relation between the true interval of orthogonality, $[\xi_1, \eta_1]$, and the coefficient sequences, $\{c_n\}$ and $\{\lambda_n\}$.

If we write

$$R_n(x) = P_n(x + s),$$

then

$$R_n(x) = (x - f_n)R_{n-1}(x) - \rho_n R_{n-2}(x), \qquad n \geq 1,$$

where

$$f_n = c_n - s, \qquad \rho_n = \lambda_n.$$

$\{R_n(x)\}$ is the monic OPS with respect to the functional $\mathfrak{M}$ defined by

$$\mathfrak{M}[x^n] = \mathcal{L}[(x - s)^n]$$

and the corresponding true interval of orthogonality is $[\xi_1 - s, \eta_1 - s]$.

Recalling the definition of a chain sequence, we see that I-Theorem 9.2 tells us that $[\xi_1 - s, \eta_1 - s] \subset [0, \infty]$ if and only if $f_n > 0$ for $n \geq 1$ and $\{\rho_{n+1}/(f_n f_{n+1})\}$ is a chain sequence. Introducing the notation

$$\alpha_n(x) = \frac{\lambda_{n+1}}{(c_n - x)(c_{n+1} - x)}, \tag{2.1}$$

we can state in summary:

THEOREM 2.1 $\xi_1 \geq s$ if and only if
  (i) $c_n > s$ $(n \geq 1)$
  (ii) $\{\alpha_n(s)\}$ is a chain sequence.

COROLLARY 1 $\eta_1 \leq t$ if and only if $c_n < t$ $(n \geq 1)$ and $\{\alpha_n(t)\}$ is a chain sequence.

Proof The polynomials $(-1)^n P_n(-x)$ satisfy (1.1) with c_n replaced by $-c_n$. The corresponding true interval of orthogonality is clearly $[-\eta_1, -\xi_1]$. By Theorem 2.1, $-\eta_1 \geqq -t$ if and only if $-c_n > -t$ $(n \geqq 1)$ and $\{\lambda_{n+1}/[(-c_n + t)(-c_{n+1} + t)]\} = \{\alpha_n(t)\}$ is a chain sequence. ∎

COROLLARY 2 $\xi_1 < c_n < \eta_1$ $(n \geqq 1)$.

THEOREM **2.2** $[\xi_1, \eta_1]$ is bounded if and only if both $\{c_n\}$ and $\{\lambda_n\}$ are bounded.

Proof If $[\xi_1, \eta_1]$ is bounded, then $\{c_n\}$ is bounded by Corollary 2. Since $\{\alpha_n(\xi_1)\}$ is a chain sequence, it is bounded and hence $\{\lambda_n\}$ is bounded.
  Conversely, if

$$A \leqq c_n \leqq B, \qquad 0 < \lambda_n \leqq C, \qquad n \geqq 1,$$

then we can choose $a < A$ such that

$$4\lambda_{n+1} \leqq 4C \leqq (c_n - a)(c_{n+1} - a), \qquad n \geqq 1.$$

We then have $c_n > a$ and $0 < \alpha_n(a) \leqq 1/4$. By the comparison test (III-Theorem 5.7), $\{\alpha_n(a)\}$ is a chain sequence, hence $a \leqq \xi_1$.
  A similar argument shows η_1 is finite. ∎

THEOREM **2.3** $[\xi_1, \eta_1] = [-\infty, \infty]$ if and only if $\{\alpha_n(x)\}$ is not a chain sequence for any real x.

Proof If $\{\alpha_n(x)\}$ is a chain sequence for some x, then since $\lambda_{n+1} > 0$, $(c_n - x)(c_{n+1} - x) > 0$ for $n \geqq 1$. Thus $c_n - x$ maintains constant sign for $n \geqq 1$. By Theorem 2.1 and Corollary 1, either ξ_1 or η_1 is finite.
  The converse is trivial. ∎
  On the basis of Theorem 2.3, each of the following is seen to be a sufficient condition for $[\xi_1, \eta_1] = [-\infty, \infty]$.

(i) $\displaystyle \inf_n c_n = -\infty$ and $\displaystyle \sup_n c_n = +\infty$;

(ii) $\{c_n\}$ is bounded while $\{\lambda_n\}$ is unbounded;   (2.2)

(iii) $\displaystyle \lim_{n\to\infty} c_n = \infty$ and $\displaystyle \liminf_{n\to\infty} \frac{\lambda_{n+1}}{c_n c_{n+1}} > \frac{1}{4}$.

Thus it is seen immediately from the recurrence formula for the Hermite polynomials (I-Ex. 1.5) that the corresponding true interval of orthogonality is $[-\infty, \infty]$. On the other hand, for the Laguerre polynomials we find $c_n = 2n + \alpha - 1$ and $\lambda_{n+1} = n(n + \alpha)$ (I-Ex. 8.5). It is not difficult to show that $\{\lambda_{n+1}/(c_n c_{n+1})\}$ is a chain sequence so the true interval of orthogonality is a subset of $[0, \infty]$. III-Theorem 5.8 can be used to show that $\{\lambda_{n+1}/[(c_n - a)(c_{n+1} - a)]\}$ is not a chain sequence if $a > 0$ so that $[\xi_1, \eta_1] = [0, \infty]$.

We now obtain an explicit formula for the minimal parameters of the chain sequence $\{\alpha_n(x)\}$, $x \notin (\xi_1, \eta_1)$.

THEOREM 2.4   Let $x \notin (\xi_1, \eta_1)$ so that $\{\alpha_n(x)\}$ is a chain sequence. Then the corresponding minimal parameter sequence $\{m_k(x)\}$ is given by

$$m_k(x) = 1 - \frac{P_{k+1}(x)}{(x - c_{k+1})P_k(x)}, \qquad k = 0, 1, 2, \ldots \qquad (2.3)$$

Proof   From (1.1) we obtain the identity

$$\frac{\lambda_{n+1}}{(x - c_n)(x - c_{n+1})} = \frac{P_n(x)}{(x - c_n)P_{n-1}(x)}\left[1 - \frac{P_{n+1}(x)}{(x - c_{n+1})P_n(x)}\right]. \quad (2.4)$$

Thus $\alpha_n(x) = [1 - m_{n-1}(x)]m_n(x)$ where $m_k(x)$ is given by (2.3).

It is clear that $m_0(x) = 0$. Thus if $\{\alpha_n(x)\}$ is a chain sequence, it then follows by induction that $\{m_k(x)\}$ is the minimal parameter sequence. ∎

As a consequence, we note that for $x \notin (\xi_1, \eta_1)$ we have

$$0 < \frac{P_{n+1}(x)}{(x - c_{n+1})P_n(x)} < 1, \qquad n \geqq 0.$$

We now obtain the result which was used in the proof of the determinacy of $\mathfrak{L}$ when $[\xi_1, \eta_1]$ is bounded (II-Theorem 5.6).

THEOREM 2.5   If $[\xi_1, \eta_1]$ is bounded,

$$\lim_{n \to \infty} p_n^2(x) = \infty \qquad \text{for } x \notin [\xi_1, \eta_1]$$

(where $p_n(x) = (\lambda_1 \lambda_2 \cdots \lambda_{n+1})^{-1/2} P_n(x)$).

Proof   From (2.3) we obtain

$$P_n^2(x) = [1 - m_0(x)]^2 \cdots [1 - m_{n-1}(x)]^2 (x - c_1)^2 \cdots (x - c_n)^2.$$

Now

$$[1 - m_{k-1}(x)](x - c_k)(x - c_{k+1}) = \frac{\lambda_{k+1}}{m_k(x)},$$

hence

$$P_n^2(x) = \frac{[1 - m_0(x)] \cdots [1 - m_{n-1}(x)](x - c_1)\lambda_2\lambda_3 \cdots \lambda_{n+1}}{m_1(x) \cdots m_{n-1}(x)m_n(x)(x - c_{n+1})}.$$

Thus for the orthonormal polynomials we can write

$$p_n^2(x) = \frac{x - c_1}{\lambda_1 m_n(x)(x - c_{n+1})\pi_n(x)}$$

where

$$\pi_n(x) = \prod_{k=1}^{n} \frac{m_k(x)}{1 - m_k(x)}.$$

Now if $x < \xi_1$, then $0 < \alpha_n(x) < \alpha_n(\xi_1)$, hence $\{\alpha_n(x)\}$ does not determine its parameters uniquely (Corollary to III-Theorem 5.5). Then by III-Theorem 6.2, $\sum \pi_n(x)$ converges, hence $\{\pi_n(x)\}$ converges to 0. Since $\{c_n\}$ is bounded, $\{p_n^2(x)\}$ diverges to ∞. ∎

Exercises

2.1 If $c_n = 0$ $(n \geq 1)$ and $\{\lambda_{n+1}\}_{n=1}^{\infty}$ is bounded, show that $[\xi_1, \eta_1] = [-a, a]$ where a is the least positive number for which $\{a^{-2}\lambda_{n+1}\}$ is a chain sequence.

2.2 Let $c_n > c$ $(n \geq 1)$ and let $\{\alpha_n(c)\}$ be a chain sequence that determines its parameters uniquely. Prove that $\xi_1 = c$.

2.3 Determine the true interval of orthogonality if (i) $c_n = n$, $\lambda_{n+1} = n$ $(n \geq 1)$; (ii) $c_{2n-1} = k_1$, $c_{2n} = k_2$, $\lambda_{n+1} = \lambda > 0$.

2.4 Use the theory of chain sequences to show that $[\xi_1^{(1)}, \eta_1^{(1)}] \subset [\xi_1, \eta_1]$ where $[\xi_1^{(1)}, \eta_1^{(1)}]$ denotes the true interval of orthogonality of the numerator OPS $\{P_n^{(1)}(x)\}$.

2.5 Suppose there is an $a > \xi_1$ such that $a < c_n$ for $n \geq 2$ and $\{\lambda_{n+2}/[(c_{n+1} - a)]\}_{n=1}^{\infty}$ · $(c_{n+2} - a)]\}_{n=1}^{\infty}$ is a chain sequence. Prove that $\xi_2 > \xi_1$ (so that $\mathcal{C}$ has a representative that has a jump at ξ_1 and is constant on (ξ_1, ξ_2)).

2.6 Show that $\{P_n(x)\}$ is itself a positive-definite KPS with K-parameter κ if and only if $c_n > \kappa$ $(n \geq 1)$ and $\{\alpha_n(\kappa)\}$ is a chain sequence that does not determine its parameters uniquely.

2.7 Using the recurrence formulas, show that $\{U_n(x)\}$ is a positive-definite KPS with K-parameter -1 while $\{T_n(x)\}$ is not. Verify by considering the corresponding weight functions.

2.8 Using the recurrence formula for the Laguerre polynomials,
   (a) verify that $\{L_n^{(\alpha)}(x)\}$ is a positive definite KPS with K-parameter $\kappa = 0$ if and only if $\alpha > 0$;
   (b) prove that $[\xi_1, \eta_1] = [0, \infty]$.

2.9 The recurrence relation for the Stieltjes-Wigert polynomials was given in I-Ex. 9.11. Show that for these polynomials, $\xi_1 \geq (1 - q)q^{-3/2} > 0$. These polynomials are known to be orthogonal with respect to a weight function which is continuous and positive on $(0, \infty)$. Conclude that the associated Hamburger moment problem is indeterminate.

2.10 Let $\{S_n(x)\}$ be the symmetric OPS related to $\{P_n(x)\}$ by the relations in I-Theorem 9.1. Use Ex. 2.2 to prove that 0 is a limit point of the set of zeros of $\{S_n(x)\}$ if

$$\sum_{n=1}^{\infty} \frac{\gamma_1 \gamma_3 \cdots \gamma_{2n-1}}{\gamma_2 \gamma_4 \cdots \gamma_{2n}} = \infty$$

2.11 (a) Show that $\frac{1}{2}(c_n + c_{n+1}) - \lambda_{n+1}^{1/2} > \xi_1$ $(n \geq 1)$.
     (b) Show that if $\{c_n - \lambda_{n+1}\}$ has a finite lower bound, then $\xi_1 > -\infty$ (Maki [4]).

2.12 Let $[\xi_1, \eta_1]$ be bounded. Prove that $\lambda_n < (\xi_1 - \eta_1)^2/4$.

2.13 Let ψ be a distribution function with an infinite spectrum which is continuous at 0 and satisfies: $\psi(x) = 0$ for $x \leq 0$, $\psi(\infty) = 1$. For any $J \geq 0$, let $\psi(x; J) = 0$ for $x < 0$ and $\psi(x; J) = J + \psi(x)$ for $x \geq 0$. Let $\{P_n(x; J)\}$ denote the monic OPS with respect to $\psi(x; J)$ and let $\{Q_n(x)\}$ denote the corresponding monic KPS with K-parameter 0. Using the notation of I-Theorem 9.1, let $\{M_n\}$ denote the maximal parameter sequence for $\{\beta_n\} = \{\nu_{n+1}/(d_n d_{n+1})\}$ (cf. Ex. 2.6).
   (a) Show that $\gamma_{2n} = h_{n-1}d_n$, $\gamma_{2n+1} = (1 - h_{n-1})d_n$ $(n \geq 1)$ where $\{h_n\}$ is any non-minimal parameter sequence for $\{\beta_n\}$.
   (b) Prove that $J = 0$ if and only if $h_0 = M_0$. [See I-Ex. 9.5 and II-Theorem 5.5.]
   (c) Show that if $J > 0$, then $h_0 = M_0/(J + 1)$.

2.14 Using the notation of I-Theorem 9.1, let $\theta > 0$ and put

$$\gamma_{2n} = \frac{n + \theta - 1}{2(n + \theta)}, \qquad \gamma_{2n+1} = \frac{n + \theta + 1}{2(n + \theta)}, \qquad n \geq 1.$$

   (a) Show that the corresponding monic KPS with K-parameter 0 is $\{Q_n(x)\}$ $= \{2^{-n}U_n(x - 1)\}$ ($U_n(x)$ is the Tchebichef polynomial of the second kind), hence that

$$P_n(x) = 2^{-n}\left[U_n(x - 1) + \frac{n + \theta + 1}{n + \theta}U_{n-1}(x - 1)\right].$$

(b) Writing $W_n(x; \alpha) = 2^n P_n(x + 1)$ and $\alpha = \theta^{-1}$, show that

$$W_{n+1}(x) = \left[2x - \frac{\alpha^2}{(\alpha n + 1)(\alpha n + \alpha + 1)} \right] W_n(x)$$

$$- \frac{(\alpha n + \alpha + 1)(\alpha n - \alpha + 1)}{(\alpha n + 1)^2} W_{n-1}(x)$$

$$W_0(x) = 1, \qquad W_1(x) = 2x + \frac{2\alpha + 1}{\alpha + 1}.$$

(c) Show that $\{W_n(x; \alpha)\}$ is an OPS with respect to

$$\psi(x; \alpha) = \begin{cases} \alpha + 1 & x > 1 \\ \alpha + \frac{1}{\pi} \int_{-1}^{x} (1 + t)^{-1/2}(1 + t)^{1/2} \, dt, & -1 < x \leq 1 \\ 0 & x \leq -1 \end{cases}$$

Note the limiting case, $\alpha = 0$.

3 Some Spectral Analysis

We turn now to a study of the relation between the coefficients in (1.1) and the limiting spectral points σ and τ. As we saw in II-Theorems 4.4 and 4.5, the limit points of the spectrum of a natural representative of $\mathcal{L}$ belong to the closed interval $[\sigma, \tau]$ (which may reduce to a single point in the extended real number system). Any ξ_i or η_j that are distinct (if any are) are isolated points of this spectrum.

In particular, if $\sigma = \tau$, this spectrum will be a denumerable set with σ as its only limit point. More specifically, this spectrum will be

$$\Xi = \{ \xi_i | i = 1, 2, 3, \ldots \} \qquad \text{if } \sigma = \tau = +\infty$$

$$H = \{ \eta_j | j = 1, 2, 3, \ldots \} \qquad \text{if } \sigma = \tau = -\infty$$

$$\Xi \cup \{ \sigma \} \cup H \qquad \text{if } -\infty < \sigma = \tau < +\infty.$$

In the latter case, $[\xi_1, \eta_1]$ is bounded so $\mathcal{L}$ is determinate.

THEOREM 3.1

$$\sigma \leq \liminf_{n \to \infty} c_n \leq \limsup_{n \to \infty} c_n \leq \tau.$$

Proof The coefficient of x^{n-1} in $P_n(x)$ is $-(c_1 + \cdots + c_n)$ (I-Theorem 4.2). By the well known relations connecting the zeros and coefficients of a polynomial,

$$\sum_{i=1}^{n} x_{ni} = \sum_{k=1}^{n} c_k. \tag{3.1}$$

Thus for any positive integer K,

$$c_n = \sum_{i=1}^{n} x_{ni} - \sum_{i=1}^{n-1} x_{n-1,i}$$

$$= x_{n1} + \sum_{i=1}^{n-1} (x_{n,i+1} - x_{n-1,i})$$

$$\geqq x_{n1} + \sum_{i=1}^{K} (x_{n,i+1} - x_{n-1,i}) \qquad \text{for } n > K.$$

Therefore

$$\liminf_{n \to \infty} c_n \geqq \xi_1 + \sum_{i=1}^{K} (\xi_{i+1} - \xi_i) = \xi_{K+1}$$

and the first inequality now follows.

Considering the polynomials $(-1)^n P_n(-x)$, we can then conclude

$$-\tau \leqq \liminf_{n \to \infty} (-c_n) = -\limsup_{n \to \infty} c_n. \quad \blacksquare$$

THEOREM 3.2   (i) If $x \notin [\sigma, \tau]$, then there is an integer $N = N(x)$ such that $\{\alpha_{N+n}(x)\}_{n=1}^{\infty}$ is a chain sequence. If $x = \sigma$ (or $x = \tau$), the same conclusion holds provided at most finitely many ξ_i are smaller than σ (η_j are larger than τ) and $c_n \leqq \sigma (c_n \leqq \tau)$ for finitely many n.

(ii) Conversely, if $\{\alpha_{N+n}(x)\}_{n=1}^{\infty}$ is a chain sequence for some integer N, then $x \notin (\sigma, \tau)$.

Proof   (i) Suppose that $x \leqq \sigma$. If $x \leqq \xi_1$, then (i) follows from Theorem 2.1, hence suppose $\xi_1 < x \leqq \sigma$. If $x < \sigma$, then there exists $p \geqq 1$ such that

$$\xi_p < x \leqq \xi_{p+1}.$$

If $x = \sigma$, then we *assume* the latter is true for some $p \geqq 1$ (so that $\xi_{p+1} = \sigma$).

Recalling (1.3), we see that there is an integer N_1 such that

$$\xi_p < x_{np} < x \leqq \xi_{p+1} < x_{n,p+1} \qquad \text{for } n \geqq N_1.$$

It then follows that for $n \geqq N$, sgn $P_n(x) = (-1)^{n-p}$ and hence $P_n(x)$ and $P_{n+1}(x)$ have opposite sign. Further, according to Theorem 3.1, $x < c_n$ for all n sufficiently large. Thus there exists an $N \geqq N_1$ such that

$$1 - m_{n-1}(x) = \frac{P_n(x)}{(x - c_n)P_{n-1}(x)} > 0, \qquad \alpha_n(x) > 0, \qquad n \geq N.$$

From the identity (2.4) we can conclude now that $m_n(x) > 0$ and hence $0 < m_n(x) < 1$ for $n \geq N$. That is,

$$\alpha_n(x) = [1 - m_{n-1}(x)]m_n(x), \qquad 0 < m_{n-1}(x) < 1, \qquad n \geq N.$$

Thus $\{\alpha_{N+n}(x)\}_{n=1}^{\infty}$ is a chain sequence (and $\{m_{N+k}(x)\}_{k=0}^{\infty}$ is a non-minimal parameter sequence).

(ii) Conversely, suppose $\{\alpha_{N+n}(x)\}_{n=1}^{\infty}$ is a chain sequence. Then $\alpha_{N+n}(x) > 0$ so $c_{N+n} - x$ does not change sign for $n \geq 1$. Suppose for definiteness that $c_{N+n} - x > 0$.

Applying Theorem 2.1 to the numerator polynomials of order N, $\{P_n^{(N)}(x)\}$, we infer that $x \leq \xi_1^{(N)}$ (where as before $[\xi_1^{(N)}, \eta_1^{(N)}]$ denotes the true interval of orthogonality for $\{P_n^{(N)}(x)\}$). The conclusion, $x \leq \sigma$, now follows from III-Theorem 4.2.

A similar argument obviously shows that $x \geq \tau$ if $c_{N+n} - x < 0$. ∎

THEOREM 3.3   For every chain sequence $\{\beta_n\}$,

$$\frac{1}{2} \liminf_{n\to\infty} \{c_n + c_{n+1} - [(c_n - c_{n+1})^2 + 4\lambda_{n+1}\beta_n^{-1}]^{1/2}\} \leq \sigma$$

$$\frac{1}{2} \limsup_{n\to\infty} \{c_n + c_{n+1} + [(c_n - c_{n+1})^2 + 4\lambda_{n+1}\beta_n^{-1}]^{1/2}\} \geq \tau.$$

Proof   Let $\{\beta_n\}$ be an arbitrary chain sequence and note that

$$c_n + c_{n+1} - [(c_n - c_{n+1})^2 + 4\lambda_{n+1}\beta_n^{-1}]^{1/2} \leq 2\min(c_n, c_{n+1}).$$

Denoting the quantity on the left side of the first inequality of the theorem by A, we can write

$$A \leq \liminf_{n\to\infty} c_n \leq \tau.$$

Now for any $x < A$, there exists an N such that for $n \geq N$,

$$2x < c_n + c_{n+1} - [(c_n - c_{n+1})^2 + 4\lambda_{n+1}\beta_n^{-1}]^{1/2}.$$

We thus have

$$(c_n - c_{n+1})^2 + 4\lambda_{n+1}\beta_n^{-1} < (c_n + c_{n+1} - 2x)^2$$

which simplifies to

$$0 < \lambda_{n+1}\beta_n^{-1} < (c_n - x)(c_{n+1} - x), \qquad n \geqq N.$$

Therefore, $0 < \alpha_n(x) < \beta_n$ for $n \geqq N$. Since $\{\beta_n^{(N)}\}$ is a chain sequence, the comparison test for chain sequences shows that $\{\alpha_{N+n}(x)\}_{n=1}^{\infty}$ is a chain sequence. Thus by Theorem 3.2, $x \notin (\sigma, \tau)$ so that $x \leqq \sigma$, and hence $A \leqq \sigma$.

The second inequality can be established similarly or it can be obtained from the first inequality by considering $\{(-1)^n P_n(-x)\}$.   ∎

We now consider special hypotheses on the coefficients in (1.1) which yield more detailed information about σ and τ.

THEOREM 3.4   Let

$$\lim_{n \to \infty} \lambda_n = 0.$$

Then

$$\sigma = \liminf_{n \to \infty} c_n, \qquad \tau = \limsup_{n \to \infty} c_n.$$

Proof   Let $\{\beta_n\}$ be any chain sequence that is bounded away from 0. Now if $\lambda_n \to 0$, then for arbitrary $\epsilon > 0$

$$[(c_n - c_{n+1})^2 + 4\lambda_{n+1}\beta_n^{-1}]^{1/2} \leqq |c_n - c_{n+1}| + \epsilon$$

for all n sufficiently large. Thus for all sufficiently large n,

$$c_n + c_{n+1} - [(c_n - c_{n+1})^2 + 4\lambda_{n+1}\beta_n^{-1}]^{1/2} \geqq 2\min(c_n, c_{n+1}) - \epsilon,$$

hence

$$\liminf_{n \to \infty}\{c_n + c_{n+1} - [(c_n - c_{n+1})^2 + 4\lambda_{n+1}\beta_n^{-1}]^{1/2}\} \geqq 2\liminf_{n \to \infty} c_n - \epsilon.$$

It follows that

$$\sigma \geqq \liminf_{n \to \infty} c_n$$

and this combined with Theorem 3.1 yields the first assertion while the second again follows by considering $\{(-1)^n P_n(-x)\}$.   ∎

Recalling now the comments at the beginning of this section regarding the significance of the case $\sigma = \tau$, we next prove:

THEOREM 3.5   A necessary and sufficient condition for $\sigma = \tau$ (finite) is

$$\lim_{n \to \infty} c_n = \sigma \quad \text{and} \quad \lim_{n \to \infty} \lambda_n = 0. \qquad (3.2)$$

Proof   Suppose $\sigma = \tau$. Then by Theorem 3.1,

$$\lim_{n \to \infty} c_n = \sigma.$$

Choose any $x < \sigma$. According to Theorem 3.2 there is an integer N such that $\{\alpha_{N+n}(x)\}_{n=0}^{\infty}$ is a chain sequence. In particular, this implies

$$0 < \alpha_n(x) < 1, \qquad n \geqq N.$$

That is,

$$0 < \lambda_{n+1} < (x - c_n)(x - c_{n+1}), \qquad n \geqq N,$$

so that

$$0 \leqq \limsup_{n \to \infty} \lambda_n \leqq (\sigma - x)^2, \qquad x < \sigma.$$

It follows that if σ is finite, then $\lim_{n \to \infty} \lambda_n = 0$. The converse is of course a special case of Theorem 3.4.   ∎

Theorem 3.5 is a generalization of a theorem of Stieltjes but is itself a special case of a theorem of M. G. Krein. Krein's theorem will be the subject of §6.

An examination of the proof of Theorem 3.4 reveals that (3.2) is a sufficient condition for $\sigma = \tau$ even if $\sigma = +\infty$. However, we will now develop more general criteria for this case. A necessary condition for $\sigma = \tau = \infty$ is of course that $\lim_{n \to \infty} c_n = \infty$.

According to Theorem 3.3, a sufficient condition for $\sigma = \infty$ is

$$\lim_{n \to \infty} \{c_n + c_{n+1} - [(c_n - c_{n+1})^2 + 4\lambda_{n+1}\beta_n^{-1}]^{1/2}\} = \infty \qquad (3.3)$$

for at least one chain sequence $\{\beta_n\}$. We first obtain a more manageable special case of (3.3).

THEOREM **3.6** A sufficient condition for $\sigma = \infty$ is $c_n \to \infty$ and

$$\lim_{n \to \infty} \frac{c_n c_{n+1} - \lambda_{n+1} \beta_n^{-1}}{c_n + c_{n+1}} = \infty. \qquad (3.4)$$

Proof We use the well known extension of Bernoulli's inequality:

$$(1 + x)^a \leqq 1 + ax \qquad \text{if } 0 < a < 1, \qquad -1 < x \leqq 0.$$

We have for n sufficiently large, $0 < 4(c_n c_{n+1} - \lambda_{n+1} \beta_n^{-1}) < (c_n + c_{n+1})^2$, hence

$$[(c_n - c_{n+1})^2 + 4\lambda_{n+1} \beta_n^{-1}]^{1/2} = |c_n + c_{n+1}| \left\{ 1 + \frac{4(\lambda_{n+1} \beta_n^{-1} - c_n c_{n+1})}{(c_n + c_{n+1})^2} \right\}^{1/2}$$

$$\leqq |c_n + c_{n+1}| \left\{ 1 + \frac{2(\lambda_{n+1} \beta_n^{-1} - c_n c_{n+1})}{(c_n + c_{n+1})^2} \right\}$$

$$= |c_n + c_{n+1}| + 2 \frac{\lambda_{n+1} \beta_n^{-1} - c_n c_{n+1}}{|c_n + c_{n+1}|}.$$

Thus if (3.4) holds, and $c_n \to \infty$, then for all sufficiently large n,

$$\frac{c_n c_{n+1} - \lambda_{n+1} \beta_n^{-1}}{c_n + c_{n+1}} \leqq \frac{1}{2} \{ c_n + c_{n+1} - [(c_n - c_{n+1})^2 + 4\lambda_{n+1} \beta_n^{-1}]^{1/2} \}.$$

Thus (3.4) implies (3.3). ∎

The simplest criterion for $\sigma = \infty$ based on Theorem 3.6 is obtained by choosing $\beta_n = 1/4$. This yields

$$\lim_{n \to \infty} \frac{c_n c_{n+1} - 4\lambda_{n+1}}{c_n + c_{n+1}} = \infty \qquad (3.5)$$

as a sufficient condition for $\sigma = \infty$.

Slightly more generally, we can take the chain sequence given by

$$\beta_n = \frac{(2n + 1)^2}{16n(n + 1)}$$

(see III-(5.2)) to obtain as a sufficient condition for $\sigma = \infty$:

$$\lim_{n \to \infty} \frac{c_n c_{n+1}}{c_n + c_{n+1}} \left\{ 1 - \frac{16n(n + 1)}{(2n + 1)^2} \frac{\lambda_{n+1}}{c_n c_{n+1}} \right\} = \infty. \qquad (3.6)$$

In particular, we have the following sufficient conditions for $\sigma = \infty$:

$$\lim_{n\to\infty} c_n = \infty \quad \text{and} \quad \limsup_{n\to\infty} \frac{\lambda_{n+1}}{c_n c_{n+1}} < \frac{1}{4} \qquad (3.7)$$

$$\lim_{n\to\infty} \frac{c_n}{n^2} = \infty \quad \text{and} \quad \frac{\lambda_{n+1}}{c_n c_{n+1}} \leq \frac{1}{4} \text{ for } n \text{ sufficiently large} . \qquad (3.8)$$

The simple criterion (3.7) is sufficiently general to apply to the majority of specific OPS that have appeared in the literature and for which $\sigma = \infty$. The preceding examples all require that

$$L = \limsup_{n\to\infty} \frac{\lambda_{n+1}}{c_n c_{n+1}} \leq \frac{1}{4} .$$

We can also find criteria that apply when $L > 1/4$. A concomitant necessary condition will be that $\liminf_{n\to\infty} \lambda_{n+1}/(c_n c_{n+1}) < 1/4$ since otherwise $\{\alpha_n(x)\}$ would not be a chain sequence for any x and we would have $\sigma = -\infty$ and $\tau = \infty$.

Writing (3.4) in the form

$$\lim_{n\to\infty} \frac{c_n c_{n+1}}{c_n + c_{n+1}} \left(1 - \frac{\alpha_n}{\beta_n} \right) = \infty, \qquad \alpha_n = \alpha_n(0),$$

we see that we will have $\sigma = \infty$ if $c_n \to \infty$ and there is a chain sequence $\{\beta_n\}$ such that

$$\limsup_{n\to\infty} \frac{\alpha_n}{\beta_n} < 1 .$$

It follows, for example, that we will have $\sigma = \infty$ if

$$\lim_{n\to\infty} c_n = \infty, \quad \limsup_{n\to\infty} \alpha_{2n-1} = A_1, \quad \limsup_{n\to\infty} \alpha_{2n} = A_2, \quad (3.9)$$

where

$$\sqrt{A_1} + \sqrt{A_2} < 1 .$$

For in this case, we can choose positive numbers a_1 and a_2 such that

$$A_1 < a_1, \quad A_2 < a_2, \quad \sqrt{a_1} + \sqrt{a_2} < 1 .$$

Then $\{\beta_n\} = \{a_1, a_2, a_1, a_2, \ldots\}$ is a periodic chain sequence (III-Theorem 6.6) and

$$\limsup_{n\to\infty} \frac{\alpha_n}{\beta_n} \leq \max\left(\frac{A_1}{a_1}, \frac{A_2}{a_2} \right) < 1 .$$

Exercises

3.1 If σ and τ are both finite, then

$$\limsup_{n\to\infty} \lambda_n \leq (\sigma - \tau)^2.$$

3.2 Prove the validity of the criteria (3.7) and (3.8).

3.3 Let $c_n = a_0 n^p + a_1 n^{p-1} + \cdots + a_p$, $a_0 > 0$, and $\lambda_n = b_0 n^{2p} + b_1 n^{2p-1} + \cdots + b_{2p} > 0$. Show that (i) if $4b_0 < a_0^2$, then $\sigma = \tau = \infty$; (ii) if $4b_0 > a_0^2$, then $\sigma = -\infty$, $\tau = \infty$.

3.4 Let $c_n = 0$ $(n \geq 1)$, $\lim_{n\to\infty} \lambda_{2n} = a$, $\lim_{n\to\infty} \lambda_{2n+1} = b$.
   (a) If $a = 0$ (if $b = 0$), the only limit points of the spectrum $\mathfrak{S}$ of a natural representative are $\pm\sqrt{b}$ (are $\pm\sqrt{a}$).
   (b) If $a = b = \infty$ and $\lim_{n\to\infty} \lambda_{2n-1}/\lambda_{2n} = L \neq 1$, then $\mathfrak{S}$ has no finite limit point.
   (c) If $a = b = \infty$ and $\lim_{n\to\infty} \lambda_{2n-1}/\lambda_{2n} = 1$ and if $\{\lambda_{2n}/\lambda_{2n-1}\}$ is a nondecreasing sequence and $\lim_{n\to\infty} n^{-2}\lambda_n = \infty$, then $\mathfrak{S}$ has no finite limit point.

3.5 The *Wall polynomials*, (Chapter VI, §11) satisty 1.1 with

$$c_{n+1} = [b + q - (1 + q)bq^n]q^n, \qquad \lambda_{n+1} = b(1 - q^n)(1 - bq^{n-1})q^{2n}$$

with $b \neq 1$, $q \neq 1$, $q > 0$. Determine σ and τ.

3.6 Prove that if there is an x and an N such that $P_n(x)$ is not zero and maintains constant sign for $n \geq N$, then $x \geq \tau$. In particular, if $N = 0$, then $x \geq \eta_1$. What is the corresponding hypothesis that yields the conclusion, $x \leq \sigma$?

4   OPS Whose Zeros are Dense in Intervals

Suppose that in (1.1),

$$\lim_{n\to\infty} c_n = c, \qquad \lim_{n\to\infty} \lambda_n = \lambda$$

where c and λ are finite. Choosing $\beta_n = 1/4$ in Theorem 3.3 then yields

$$c - 2\sqrt{\lambda} \leq \sigma, \qquad \tau \leq c + 2\sqrt{\lambda}. \tag{4.1}$$

Further, if $|x_0 - c| < 2\sqrt{\lambda}$, then

$$\lim_{n\to\infty} \alpha_n(x_0) = \frac{\lambda}{(x_0 - c)^2} > \frac{1}{4},$$

hence according to III-Theorem 5.6, Corollary 2, $\{\alpha_{N+n}(x_0)\}_{n=1}^{\infty}$ cannot be a chain sequence for any N. It then follows from Theorem 3.2 that

$x_0 \in [\sigma, \tau]$. This combined with (4.1) then yields the conclusion

$$\sigma = c - 2\sqrt{\lambda}, \qquad \tau = c + 2\sqrt{\lambda}. \tag{4.2}$$

In 1898, O. Blumenthal proved that the set X of all zeros of all polynomials $P_n(x)$ is dense in the interval $[\sigma, \tau]$. We will prove Blumenthal's theorem as well as some extensions of it. However, for these proofs, we will have to reach beyond the elementary prerequisites we have assumed up to now. Specifically, we will need to use the following classical theorem from complex function theory.

The Stieltjes-Vitali Theorem: Let $\{f_n\}$ be a sequence of analytic functions, each regular in an open region G of the complex plane. If $\{f_n\}$ is uniformly bounded on G and converges on a subset E of G where E has a limit point in G, then $\{f_n\}$ converges uniformly on G.

We will assume the Stieltjes-Vitali theorem without proof. For a proof as well as interesting historical remarks, see Hille [1]. It is interesting to observe that the initial form of this theorem was proved by Stieltjes in the same classic memoir [2] in which he solved his moment problem and introduced the Stieltjes integral.

Introduce now the rational function

$$F_n(z) = \frac{(z - c_{n+1}) P_n(z)}{P_{n+1}(z)}, \tag{4.3}$$

which should be compared with (2.3). Writing

$$X = \{x_{ni} | 1 \leq i \leq n; \; n = 1, 2, 3, \dots\}$$

we obtain an inequality involving $F_n(z)$.

LEMMA Let $\xi_1 > -\infty$ and let E be any bounded subset of the complex plane which is a positive distance from X. Then for every real $x_0 < \xi_1$, there is a corresponding constant $K_0 > 0$ such that

$$|F_n(z)| \leq K_0 |F_n(x_0)|, \qquad z \in E. \tag{4.4}$$

Proof Let δ denote the distance between E and X. That is, put

$$\delta = \inf |z - x_{nk}| \qquad (z \in E, x_{nk} \in X).$$

By hypothesis, $\delta > 0$. From I-(5.8) and (5.9), we have for $z \in E$,

$$|F_n(z)| \leqq K \left| \frac{z - c_{n+1}}{x_0 - c_{n+1}} \right| |F_n(x_0)| \leqq K \left(1 + \frac{|x_0 - z|}{|x_0 - c_{n+1}|} \right) |F_n(x_0)|$$

where

$$K = \max_{1 \leqq k \leqq n} \left| \frac{x_0 - x_{nk}}{z - x_{nk}} \right| \leqq 1 + \frac{|x_0 - z|}{\delta}.$$

Since $|x_0 - c_{n+1}| > \xi_1 - x_0 > 0$, we obtain (4.4) with

$$K_0 = \sup_{z \in E} \left(1 + \frac{|x_0 - z|}{\delta} \right) \left(1 + \frac{|x_0 - z|}{\xi_1 - x_0} \right).$$

By hypothesis, E is bounded so K_0 is finite. ∎

We can now prove the following generalization of Blumenthal's theorem.

THEOREM 4.1   Let

$$\lim_{n \to \infty} c_{2n-1} = k_1, \qquad \lim_{n \to \infty} c_{2n} = k_2, \qquad \lim_{n \to \infty} \lambda_n = \lambda.$$

Then

$$\sigma = \frac{1}{2}\{k_1 + k_2 - [(k_1 - k_2)^2 + 16\lambda]^{1/2}\}$$

$$\tau = \frac{1}{2}\{k_1 + k_2 + [(k_1 - k_2)^2 + 16\lambda]^{1/2}\}$$

$$(4.5)$$

and X is dense in $[\sigma, \sigma^*] \cup [\tau^*, \tau]$ where

$$\sigma^* = \min(k_1, k_2), \qquad \tau^* = \max(k_1, k_2).$$

Proof   Taking $\beta_n = 1/4$ in Theorem 3.3, we obtain

$$A = \frac{1}{2}\{k_1 + k_2 - [(k_1 - k_2)^2 + 16\lambda]^{1/2}\} \leqq \sigma$$

$$B = \frac{1}{2}\{k_1 + k_2 + [(k_1 + k_2)^2 + 16\lambda]^{1/2}\} \geqq \tau.$$

Now

$$\lim_{n \to \infty} \alpha_n(x) = L(x) \equiv \frac{\lambda}{(x - k_1)(x - k_2)}$$

and routine calculation shows that $0 \leq L(x) \leq 1/4$ if and only if $x \leq A$ or $x \geq B$. Since the limit, L, of a convergent chain sequence must satisfy $0 \leq L \leq 1/4$ (III-Theorem 6.4), it follows that $\{\alpha_{N+n}(x)\}$ cannot be a chain sequence for any N if $A < x < B$. It then follows from Theorem 3.2 that if $A < x < B$, then $\sigma \leq x \leq \tau$. That is,

$$\sigma \leq A \leq B \leq \tau.$$

This establishes (4.5).

Next put $\sigma^* = \min(k_1, k_2)$ and assume $[\sigma, \sigma^*]$ contains a subinterval that is free of the zeros x_{ni}. Then there is an open interval $(\alpha, \beta) \subset [\sigma, \sigma^*]$ which is a positive distance from X. Hence we can choose a bounded, open region G which (i) is a positive distance from X; (ii) contains (α, β); and (iii) contains a segment of the interval $(-\infty, \xi_1)$ (figure 1). Let $E = G \cap (-\infty, \xi_1)$.

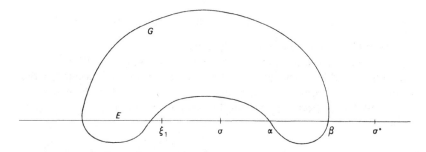

Figure 1

Now for $x \leq \xi_1$, $\{\alpha_n(x)\}$ is a chain sequence whose minimal parameters are given by Theorem 2.4 as

$$m_n(x) = 1 - [F_n(x)]^{-1}, \qquad n \geq 0.$$

Further, for $x < \xi_1 \leq \sigma$, $0 \leq L(x) < 1/4$ so by III-Theorem 6.4,

$$\lim_{n \to \infty} m_n(x) = \frac{1}{2}\{1 - [1 - 4L(x)]^{1/2}\} < \frac{1}{2}.$$

It follows that $\{F_n(x)\}$ converges for $x \in E$. By the lemma, $\{F_n\}$ is uniformly bounded on G, hence by the Stieltjes-Vitali theorem it converges uniformly on G. In particular, $\{F_n(x)\}$ converges for $x \in (\alpha, \beta)$ to some $F(x)$ which must be real since each $F_n(x)$ is real.

But from the recurrence formula, we must have

$$\frac{1}{F(x)}\left[1 - \frac{1}{F(x)}\right] = \frac{\lambda}{(x - k_1)(x - k_2)}, \qquad x \in (\alpha, \beta).$$

Now for real $a = [F(x)]^{-1}$, $a(1 - a) \leq 1/4$, while if $\sigma < x < \sigma^*$,

$$\frac{\lambda}{(x - k_1)(x - k_2)} > \frac{1}{4}.$$

This contradiction establishes the theorem. ∎

We note that if $\sigma^* < x < \tau^*$, then $L(x) < 0$ and no contradiction is arrived at in the above proof if $\{F_n(x)\}$ converges for $\sigma^* < x < \tau^*$.

The case $k_1 = k_2 = c$ is of course Blumenthal's theorem. (Blumenthal actually assumes also that $\lambda > 0$.) In practically all of the examples of this type occuring in the literature, it is the case that

$$\xi_1 = \sigma = c - 2\sqrt{\lambda}, \qquad \eta_1 = \tau = c + 2\sqrt{\lambda}.$$

For example, note the Legendre polynomials.

However, it is not difficult to construct (see Ex. 4.6; also, VI–§13 C, L) examples in which there are a finite number of spectral points on the complement of $[\sigma, \tau]$. The following example shows that there can be countably many spectral points on the complement of $[\sigma, \tau]$.

Let

$$c_n = c, \qquad \lambda_{n+1} = \lambda(1 + \frac{1}{n}), \qquad \lambda > 0, \quad n \geq 1.$$

The conditions of Blumenthal's theorem are satisfied so $\sigma = c - 2\sqrt{\lambda}$. However,

$$\alpha_n(\sigma) = \frac{1}{4}(1 + \frac{1}{n}),$$

hence by III-Theorem 5.8, Corollary, $\{\alpha_{N+n}(\sigma)\}$ is not a chain sequence for any N. From Theorem 3.2 it follows that $\xi_i < \sigma$ for all i.

No explicit formulas are known for the orthogonal polynomials of the last example nor is the orthogonality relation known. There does not seem to be any example studied in the literature which exhibits the sort of "spectral property" described in this example.

Our next theorem provides an analogue of Blumenthal's theorem for the case where (1.1) has unbounded coefficients.

THEOREM **4.2** Let

$$\lim_{n\to\infty} c_n = \infty, \qquad \lim_{n\to\infty} \frac{\lambda_{n+1}}{c_n c_{n+1}} = \frac{1}{4}.$$

Then if σ is finite, X is dense in $[\sigma, \infty)$.

Proof The proof is very similar to that of Theorem 4.1. Assume that $[\sigma, \infty)$ contains a subinterval that is free of zeros so that there exists an open interval $(\alpha, \beta) \subset [\sigma, \infty)$ which is a positive distance from X.

Now if σ is finite, so is ξ_1 so there is a bounded region G which is a positive distance from X, contains (α, β), and contains a subinterval of $(-\infty, \xi_1)$. Again put $E = G \cap (-\infty, \xi_1)$.

As before, $\{\alpha_n(x)\}$ is a chain sequence for $x \leq \xi_1$. Further, $\alpha_n(x) \to 1/4$ $(n \to \infty)$ so that

$$\lim_{n\to\infty} m_n(x) = \frac{1}{2}, \qquad \lim_{n\to\infty} F_n(x) = 2, \qquad x \leq \xi_1.$$

Thus $\{F_n\}$ converges on E, hence by the Stieltjes-Vitali theorem we again conclude that $\{F_n\}$ converges uniformly on G to a function F which is regular in G. By the identity theorem for analytic functions, $F(z) = 2$ for all $z \in G$. This means, in particular, that for $x \in (\alpha, \beta)$, there exists an N such that $F_n(x) > 1$ for $n \geq N$. That is,

$$0 < m_n(x) = 1 - \frac{1}{F_n(x)} < 1 \qquad \text{for } n \geq N, \qquad \alpha < x < \beta.$$

But $\alpha_n(x) = [1 - m_{n-1}(x)]m_n(x)$ is an identity for all $x \notin X$, hence $\{\alpha_{N+n}(x)\}_{n=1}^{\infty}$ must be a chain sequence for $\alpha < x < \beta$. This contradicts Theorem 3.2. ■

The fact that the finiteness of σ must be assumed as a hypothesis and does not follow from the other conditions of Theorem 4.2 is shown by the following example. Let

$$c_n = 2n, \qquad \lambda_{n+1} = n^2 + n^\gamma, \qquad 1 < \gamma < 2.$$

We can write, for any real x,

$$\alpha_n(x) = \frac{1}{4} + r_n(x)$$

where

$$r_n(x) = \frac{4n^\gamma - 4n(1-x) - (x^2 - 2x)}{16n^2 + 16n(1-x) + 4(x^2 - 2x)}.$$

It follows that there is an $N = N(x)$ such that $r_n(x) > 0$ for $n \geqq N$ and $\sum_{n=N}^{\infty} r_n(x) = \infty$. By III-Theorem 5.8, Corollary, $\{\alpha_n(x)\}$ is not a chain sequence for any real x. Thus $\xi_1 = \sigma = -\infty$.

Of course, nothing precludes the possibility that the corresponding set of zeros is dense in $(-\infty, \infty)$ in the latter example. We know of no specific examples in which $\sigma = -\infty$ and the conditions on $\{c_n\}$ and $\{\lambda_n\}$ of Theorem 4.2 are fulfilled.

On the other hand, there is a known example in which $\sigma = +\infty$ but the remaining hypotheses of Theorem 4.2 are satisfied. (Of course, in such a case the conclusion of Theorem 4.2 is vacuously true.) The polynomials satisfying (1.1) with

$$c_{n+1} = 2q^{-n}, \qquad \lambda_{n+1} = q^{1-n}(q^{-n} - 1), \qquad 0 < q < 1,$$

are a special case of a class of orthogonal polynomials studied by W. A. Al-Salam and L. Carlitz. They have shown that these polynomials are orthogonal with respect to a distribution function whose spectrum consists of the points $q^{-k}(k = 0, 1, 2, \ldots)$. These polynomials are described briefly in Chapter VI.

An interesting illustration of Theorem 4.2 is provided by taking

$$c_n = an + b, \qquad a > 0, \qquad \lambda_n = dn^2 + fn + g,$$

where d, f, and g must be restricted so that $\lambda_n > 0$ for $n \geqq 2$. In particular, this requires that $d > 0$.

Since

$$\lim_{n \to \infty} \frac{\lambda_{n+1}}{c_n c_{n+1}} = \frac{d}{a^2},$$

it follows that

     (i) if $4d < a^2$, then $\sigma = \infty$ (by (3.7))

     (ii) if $4d > a^2$, then $\sigma = -\infty$ (by (2.2)).

     (iii) If $4d = a^2$, rewrite $\alpha_n(x)$ in the form

$$\alpha_n(x) = \frac{1}{4} - \frac{A(x)n + B(x)}{4dn^2 + C(x)n + D(x)}.$$

where

$$A(x) = (b - d^{1/2} - fd^{-1/2} - x)d^{1/2}$$

and $B(x)$, $C(x)$, and $D(x)$ are independent of n.

Then for $x < b - d^{1/2} - fd^{-1/2}$, $A(x) > 0$ so $0 < \alpha_n(x) < 1/4$ for n sufficiently large. That is, for N sufficiently large, $\{\alpha_{N+n}(x)\}$ is a chain sequence. Therefore by Theorem 3.2, $x \leqq \sigma$, hence

$$b - d^{1/2} - fd^{-1/2} \leqq \sigma.$$

On the other hand, for $x > b - d^{1/2} - fd^{-1/2}$, for N sufficiently large, $\alpha_n(x) > 1/4$ $(n \geqq N)$ and

$$\sum_{n=N}^{\infty} \left[\alpha_n(x) - \frac{1}{4} \right] = \infty.$$

Thus by III-Theorem 5.8, Corollary, $\{\alpha_{N+n}(x)\}$ is not a chain sequence for any N. Once again calling upon Theorem 3.2, we conclude $x \geqq \sigma$, hence $b - d^{1/2} - fd^{-1/2} \geqq \sigma$.

We can now conclude that

$$\sigma = b - d^{1/2} - fd^{-1/2}.$$

According to Theorem 4.2, the zeros of the corresponding orthogonal polynomials are dense in $[\sigma, \infty)$.

Exercises

4.1 Using only the recurrence formula, discuss the density of the zeros of the
    (i) Legendre polynomials (I-Ex. 1.6)
    (ii) Laguerre polynomials (I-Ex. 8.5)
    (iii) Hermite polynomials (I-Ex. 1.5)
    (iv) Charlier polynomials (I-Ex. 4.10)
    (v) Stieltjes-Wigert polynomials (I-Ex. 9.11).

4.2 Let $c_n = 0$ $(n \geqq 1)$ and $\lim_{n\to\infty} \lambda_n = 1/4$. Prove that if $\{\lambda_{n+1}\}$ is a chain sequence, then $\mathfrak{S}(\psi) = [-1, 1]$, where ψ denotes the substantially unique representative for the corresponding moment functional.

4.3 Let $c_n = 0$ $(n \geqq 1)$, $\lim_{n\to\infty} \lambda_{2n} = a$, $\lim_{n\to\infty} \lambda_{2n+1} = b$ (both finite). Show that the zeros of $\{P_n(x)\}$ are dense in $[-\beta, -\alpha] \cup [\alpha, \beta]$ where $\alpha = |\sqrt{a} - \sqrt{b}|$, $\beta = \sqrt{a} + \sqrt{b}$.

4.4 Let $c_n = 0$ $(n \geqq 1)$ and let

$$\lambda_{2n} = \frac{(n+1)(2n+\alpha-1)}{n}, \qquad \lambda_{2n+1} = \frac{n(2n+\alpha+1)}{n+1}, \qquad \alpha > -1.$$

Prove that the zeros of $\{P_n(x)\}$ are dense in $(-\infty, \infty)$.

4.5 Prove that if $\lim_{n\to\infty} a_n(x) = \alpha(x)$ where $0 \leqq \alpha(x) \leqq 1/4$, then

$$\lim_{n\to\infty} \frac{P_{n+1}(x)}{(x - c_{n+1})P_n(x)} = \frac{1}{2}[1 \pm \sqrt{1 - 4\alpha(x)}\,].$$

4.6 Let $U_n(x,c)$ be defined by

$$U_n(x) = xU_{n-1}(x) - \frac{1}{4}U_{n-2}(x), \qquad n \geqq 2,$$

$$U_0(x) = 1, \qquad U_1(x) = x - c, \qquad U_n(x) = U_n(x,c).$$

Then $\sigma = -1$ and $\tau = 1$. Show that $\xi_1 < \sigma$ if $c > 1/2$. However, $\xi_2 = \sigma$.

4.7 If $c_n = 2n + p - 1$, $\lambda_{n+1} = n^2 + pn + q > 0$ $(n \geqq 1)$, the corresponding orthogonality relations are not known for $q \neq 0$. Show however that if the equation $t^2 - pt + q = 0$ has a root α such that $0 \leqq \alpha < p + 1$, then the spectrum of the corresponding distribution function is $[0, \infty)$. Assume the associated moment problem is determined. (These polynomials have been studied briefly by Carlitz [8].)

5 Preliminaries to Krein's Theorem

Our final objective in this portion of the book is to develop the important theorem of Krein which provides a characterization of the case where $\mathfrak{S}(\psi)$ is a bounded set with finitely many limit points. As mentioned earlier, this result contains Theorem 3.5 as a special case.

We first review some basic facts about the space of real continuous functions on a compact interval viewed, essentially, as an incomplete inner product space. The notation introduced previously in this chapter will be maintained and it will be assumed throughout the remainder of this chapter that $\{c_n\}$ and $\{\lambda_n\}$ are bounded sequences. Thus the associated distribution function ψ is substantially unique and its spectrum $\mathfrak{S}(\psi)$ is a uniquely determined compact set.

Let us write

$$E = \mathfrak{S}(\psi), \qquad J = [\xi_1, \eta_1]$$

and denote by $C(J)$ the space of all real functions continuous on J. With the usual definitions of addition and multiplication by real scalars, $C(J)$ becomes a vector space over the real numbers.

Now define

$$(f,g) = \int_J f(x)g(x)\,d\psi(x), \qquad f, g \in C(J) \tag{5.1}$$

We have the easily verified properties:

$(f,f) \geqq 0$ with $(f,f) = 0$ if and only if $f(x) = 0$ for all $x \in E$;

$(f,g) = (g,f)$;

$(\alpha f + \beta g, h) = \alpha(f,h) + \beta(g,h)$;

for all f, g, $h \in C(J)$ and real numbers α and β. (Thus (5.1) would define an inner product if we dealt with equivalence classes of functions which differ only on the complement of E. We will not need this refinement for our work however).

Next writing

$$\|f\| = (f,f)^{1/2} \geqq 0,$$

we have for any real α,

$$0 \leqq \|\alpha f + g\|^2 = (\alpha f + g, \alpha f + g)$$
$$= \alpha^2(f,f) + 2\alpha(f,g) + (g,g).$$

Thus

$$0 \leqq \|f\|^2 \alpha^2 + 2(f,g)\alpha + \|g\|^2$$

so, as a quadratic in α, the right side must have a non-positive discriminant:

$$4(f,g)^2 - 4\|f\|^2 \|g\|^2 \leqq 0.$$

We therefore obtain *Schwarz' inequality*,

$$|(f,g)| \leqq \|f\| \|g\|.$$

The latter can be used to prove the *triangle inequality*,

$$\|f + g\| \leqq \|f\| + \|g\|,$$

since

$$\|f + g\|^2 = \|f\|^2 + 2(f,g) + \|g\|^2$$
$$\leqq \|f\|^2 + 2\|f\| \|g\| + \|g\|^2 = (\|f\| + \|g\|)^2.$$

Now consider the orthonormal polynomials, $p_n(x)$, associated with ψ:

$$p_n(x) = (\lambda_1 \lambda_2 \cdots \lambda_{n+1})^{1/2} P_n(x)$$
$$(p_m, p_n) = \delta_{mn}, \qquad m, n = 0, 1, 2, \ldots.$$

For any $f \in C(J)$, we have the associated formal *Fourier expansion*,

$$f \sim \sum_{k=0}^{\infty} c_k p_k, \qquad c_k = (f, p_k). \tag{5.2}$$

The series in (5.2) does not in general converge pointwise to f. However we do have:

THEOREM 5.1   Let $f \in C(J)$, $c_k = (f, p_k)$. Then

$$\lim_{n \to \infty} \left\| f - \sum_{k=0}^{n} c_k p_k \right\| = 0. \tag{5.3}$$

Proof   Since J is compact, the Weierstrass approximation theorem assures us there are polynomials $\pi_n(x)$ such that

$$\lim_{n \to \infty} |f(x) - \pi_n(x)| = 0 \text{ uniformly on } J. \tag{5.4}$$

We can always choose the polynomials so that $\pi_n(x)$ has degree at most n so we write

$$\pi_n(x) = \sum_{k=0}^{n} \gamma_{nk} p_k(x).$$

We then have

$$\|f - \pi_n\|^2 = \|f\|^2 - 2 \sum_{k=0}^{n} \gamma_{nk}(f, p_k) + \sum_{i,j=0}^{n} \gamma_{ni} \gamma_{nj} (p_i, p_j)$$

$$= \|f\|^2 - 2 \sum_{k=0}^{n} \gamma_{nk} c_k + \sum_{i=0}^{n} \gamma_{ni}^2,$$

and

$$\left\| f - \sum_{k=0}^{n} c_k p_k \right\|^2 = \|f\|^2 - \sum_{k=0}^{n} c_k^2$$

$$= \|f - \pi_n\|^2 - \sum_{k=0}^{n} (c_k - \gamma_{nk})^2 \leqq \|f - \pi_n\|^2.$$

But

$$\|f - \pi_n\|^2 \leqq \int_J |f(x) - \pi_n(x)|^2 \, d\psi(x) \leqq \mu_0 \max_{t \in J} |f(t) - \pi_n(t)|^2$$

so that (5.4) implies (5.3).   ∎

Henceforth, to distinguish the "mean convergence" of (5.3) from ordinary pointwise convergence of the series in (5.2), we write for any sequence $\{\phi_n\}$ in $C(J)$,

$$f \doteq \sum_{k=0}^{\infty} \phi_k$$

to mean

$$\lim_{n \to \infty} \left\| f - \sum_{k=0}^{n} \phi_k \right\| = 0.$$

THEOREM **5.2** Let

$$f \doteq \sum_{k=0}^{\infty} a_k p_k$$

Then
  (i) $a_k = (f,p_k)$, $k = 0, 1, 2, \ldots.$
  Moreover, if

$$g \doteq \sum_{k=0}^{\infty} b_k p_k$$

then
  (ii) $\alpha f + \beta g \doteq \sum_{k=0}^{\infty} (\alpha a_k + \beta b_k) p_k$;
  (iii) $(f,g) = \sum_{k=0}^{\infty} a_k b_k$.

Proof Set

$$f_n = \sum_{k=0}^{n} a_k p_k$$

so that $a_k = (f_n, p_k)$ (I-Theorem 2.2 applies since f_n is a polynomial). By Schwarz' inequality,

$$|(f,p_k) - a_k| = |(f - f_n, p_k)| \leq \| f - f_n \|.$$

Therefore, if $\| f - f_n \| \to 0$, it follows that $a_k = (f,p_k)$.
  It is now trivial to conclude (ii). As for (iii), we have

$$|(f,g) - (f_n,g)| \leq \| f - f_n \| \| g \|$$

so that

$$(f,g) = \lim_{n \to \infty} (f_n,g) = \lim_{n \to \infty} \sum_{k=0}^{n} a_k (p_k, g) = \lim_{n \to \infty} \sum_{k=0}^{n} a_k b_k. \quad \blacksquare$$

An immediate consequence of (iii) is *Parseval's identity* which is valid for every $f \in C(J)$:

$$\|f\|^2 = \sum_{k=0}^{\infty} c_k^2, \qquad c_k = (f, p_k).$$

THEOREM **5.3**   Let I be a subinterval of J such that $E^* = E \cap I$ is an infinite set. Then there is a sequence $\{\phi_n\}_{n=1}^{\infty}$ in $C(J)$ such that
   (i) $\{\phi_n\}$ is orthonormal: $(\phi_m, \phi_n) = \delta_{mn}$;
   (ii) $\phi_n(x) = 0$ for $x \notin I$ $(n \geq 1)$.

Proof   Since E^* is an infinite set, we can construct a sequence $\{f_n\}$ of continuous functions each of which vanishes outside I and such that, for each n, $f_1, f_2, \ldots, f_n$ are linearly independent on E^*.
   Then

$$Q_n = \sum_{i,j=1}^{n} (f_i, f_j) y_i y_j = \int_J \sum_{i,j=1}^{n} y_i y_j f_i(x) f_j(x) \, d\psi(x)$$

$$= \int_I \left[\sum_{k=1}^{n} y_k f_k(x) \right]^2 d\psi(x) > 0$$

unless all $y_i = 0$. That is, Q_n is a positive-definite quadratic form, hence it has a positive determinant:

$$\det((f_i, f_j))_{i,j=1}^{n} > 0 \qquad (n = 1, 2, 3, \ldots).$$

If we now let

$$g_n = \sum_{i=1}^{n} c_{ni} f_i,$$

it follows that the system of equations

$$(g_n, f_k) = \sum_{i=1}^{n} c_{ni}(f_i, f_k) = \delta_{nk}, \qquad k = 1, 2, \ldots, n$$

has a unique solution. Thus we can determine the c_{ni} so that $(g_m, g_n) = 0$ for $m \neq n$, $\|g_n\| > 0$. Then $\phi_n = g_n / \|g_n\|$ provides the orthonormal functions we want. ■

THEOREM **5.4**   Let D be a finite or denumerable set of isolated points of E and let $\{f_n\}$ be any sequence in $C(J)$ with bounded norms: $\|f_n\| \leq M$ $(n \geq 1)$. Then $\{f_n\}$ contains a subsequence which converges pointwise on D.

Proof Let N denote the cardinality of D if D is finite and write $N = \infty$ otherwise. Writing $D = \{y_1, y_2, \ldots\}$, we observe that since each y_i is an isolated point of E, ψ has a positive jump at y_i:

$$\rho_i = \psi(y_i + 0) - \psi(y_i - 0) > 0.$$

It follows that

$$|f_n(y_k)|^2 \leqq \frac{1}{\rho_k} \sum_{i=1}^{N} f_n^2(y_i)\rho_i \leqq \frac{1}{\rho_k} \int_J f^2(y)\,d\psi(y) \leqq \frac{M^2}{\rho_k}.$$

Thus for each k, $\{f_n(y_k)\}_{n=1}^{\infty}$ is a bounded sequence and the desired conclusion follows from II-Theorem 2.1. ∎

Exercises

5.1 Prove the validity of the "parallelogram law":

$$\|f + g\|^2 + \|f - g\|^2 = 2(\|f\|^2 + \|g\|^2).$$

5.2 When does equality hold in Schwarz' inequality?

5.3 Verify the remark made in the proof of Theorem 5.1 that the polynomials, $\pi_n(x)$, provided by the Weierstrass approximation theorem can be chosen so that the degree of $\pi_n(x)$ does not exceed n.

5.4 Let E be an arbitrary, infinite bounded set of real numbers. Construct explicitly an example of a sequence of functions f_n with the properties described in the opening sentence of the proof of Theorem 5.3.

5.5 Let $E = \mathfrak{S}(\psi)$ be a bounded set with a single limit point σ. Let $\{f_n\}$ be any sequence in $C(J)$ such that $\|f_n\| \leqq M$ $(n \geqq 1)$ and let $g_n(x) = (x - \sigma)f_n(x)$. Prove there is a subsequence $\{g_{n_k}\}$ such that $\lim_{j,k\to\infty}\|g_{n_j} - g_{n_k}\| = 0$. (The operator A, $Af(x) = (x - \sigma)f(x)$, is completely continuous.)

6 Krein's Theorem

Throughout this section, $\pi(x)$ denotes a polynomial of degree p and we write

$$\pi(x)p_n(x) = \sum_{k=0}^{\infty} \pi_{nk} p_k(x) \tag{6.1}$$

so that

$$\pi_{nk} = (\pi p_n, p_k) = (p_n, \pi p_k) = \pi_{kn}. \tag{6.2}$$

It also follows from the orthogonality properties of $\{p_n\}$ that

$$\pi_{nk} = 0 \qquad \text{if } |n - k| > p. \tag{6.3}$$

(It will be convenient if we define $\pi_{nk} = 0$ for $k < 0$).

Further note that

$$\|\pi p_m - \pi p_n\|^2 = \sum_{k=0}^{\infty} \pi_{mk}^2 - 2 \sum_{k=0}^{\infty} \pi_{mk} \pi_{nk} + \sum_{k=0}^{\infty} \pi_{nk}^2.$$

If $|m - n| > 2p$, then

$$|m - k| + |n - k| \geqq |m - n| > 2p$$

so that $\pi_{mk} \pi_{nk} = 0$. Thus we have

$$\|\pi p_m - \pi p_n\|^2 = \sum_{k=m-p}^{m+p} \pi_{mk}^2 + \sum_{k=n-p}^{n+p} \pi_{nk}^2 \qquad \text{if } |m - n| > 2p. \tag{6.4}$$

LEMMA   Let $\phi \in C(J)$ and let

$$\pi\phi \doteq \sum_{k=0}^{\infty} d_k p_k.$$

If

$$|\pi_{nk}| \leqq \epsilon \qquad \text{for } k \geqq K, \qquad n = 0, 1, 2, \ldots,$$

then

$$\sum_{k=K}^{\infty} d_k^2 \leqq [(2p + 1)\epsilon\|\phi\|]^2.$$

Proof   Let

$$\phi \doteq \sum_{k=0}^{\infty} c_k p_k$$

and conclude from Theorems 5.1 and 5.2,

$$d_k = (\pi\phi, p_k) = (\phi, \pi p_k) = \sum_{j=0}^{\infty} c_j \pi_{jk},$$

$$d_k = \sum_{j=k-p}^{k+p} c_j \pi_{jk}. \tag{6.5}$$

Using Cauchy's inequality, we then have for $k \geqq K$,

$$d_k^2 \leqq \sum_{j=k-p}^{k+p} c_j^2 \sum_{j=k-p}^{k+p} \pi_{jk}^2 \leqq (2p + 1)\epsilon^2 \sum_{j=k-p}^{k+p} c_j^2.$$

Therefore,

$$\sum_{k=K}^{\infty} d_k^2 \leqq (2p + 1)\epsilon^2 \sum_{k=K}^{\infty} \sum_{j=k-p}^{k+p} c_j^2 = (2p + 1)\epsilon^2 \sum_{k=K-p}^{K+p} \sum_{i=k}^{\infty} c_i^2$$

$$\leqq (2p + 1)^2 \epsilon^2 \|\phi\|^2. \quad \blacksquare$$

Finally, we are ready to prove Krein's Theorem.

THEOREM 6.1   Every limit point of E is a zero of $\pi(x)$ if and only if

$$\lim_{n \to \infty} \pi_{n,n+i} = 0 \qquad \text{for } i = 0, 1, \ldots, p. \tag{6.6}$$

Proof   Assume first that (6.6) fails for $i = k$, $0 \leqq k \leqq p$. That is, assume there is a $\delta > 0$ and an increasing sequence of positive integers n_i such that

$$\pi_{n_i,n_i+k} \geqq \delta > 0 \qquad (i = 1, 2, 3, \ldots).$$

We can clearly choose this sequence so that $n_{i+1} > n_i + 2p$. With this additional condition met, we then have by (6.4),

$$\left\| \pi p_{n_i} - \pi p_{n_j} \right\|^2 = \sum_{\nu} \pi_{n_i \nu}^2 + \sum_{\nu} \pi_{n_j \nu}^2 \geqq 2\delta^2. \tag{6.7}$$

Now suppose every limit point of E is a zero of $\pi(x)$. We can write $E = E_1 \cup E_2$ where $\pi(E_1) = \{0\}$ and $E_2 = \{y_1, y_2, y_3, \ldots\}$ with $\pi(y_i) \neq 0$.

Let $\{\phi_i\}_{i=1}^{\infty}$ be any sequence in $C(J)$ with $\|\phi_n\| \leqq M$. By Theorem 5.4, there is a subsequence $\{\phi_{i_s}\}_{s=1}^{\infty}$ which converges on E_2. Since the limit points of E_2 are zeros of $\pi(x)$, there is an integer N such that

$$|\pi(y_k)| < \frac{\delta}{2M} \qquad \text{for } k > N.$$

Therefore

$$\left\| \pi\phi_{i_s} - \pi\phi_{i_t} \right\|^2 = \int_J \pi^2(y) \left[\phi_{i_s}(y) - \phi_{i_t}(y) \right]^2 d\psi(y)$$

$$= \sum_{k=1}^{\infty} \pi^2(y_k) \left[\phi_{i_s}(y_k) - \phi_{i_t}(y_k) \right]^2 \rho_k$$

where $\rho_k = \psi(y_k + 0) - \psi(y_k - 0) > 0$. Further,

$$\sum_{k=N+1}^{\infty} \pi^2(y_k)\left[\phi_{i_s}(y_k) - \phi_{i_t}(y_k)\right]^2 \rho_k$$

$$< \frac{\delta^2}{4M^2} \sum_{k=N+1}^{\infty} \left[\phi_{i_s}(y_k) - \phi_{i_t}(y_k)\right]^2 \rho_k \leq \frac{\delta^2}{4M^2}\left\|\phi_{i_s} - \phi_{i_t}\right\|^2 \leq \delta^2,$$

hence

$$\left\|\pi\phi_{i_s} - \pi\phi_{i_t}\right\|^2 \leq \sum_{k=1}^{N} \pi^2(y_k)\left[\phi_{i_s}(y_k) - \phi_{i_t}(y_k)\right]^2 \rho_k + \delta^2.$$

Since $\{\phi_{i_s}(y_k)\}_{s=1}^{\infty}$ converges for $k = 1, 2, \ldots, N$, we conclude

$$\limsup_{s,t \to \infty} \left\|\pi\phi_{i_s} - \pi\phi_{i_t}\right\|^2 \leq \delta^2.$$

Taking $\phi_i = p_{n_i}$, we obtain a contradiction of (6.7), That is, if every limit point of E is a zero of $\pi(x)$, then (6.6) must hold.

Conversely, suppose we have (6.6) and assume E has a limit point ξ where $\pi(\xi) \neq 0$. Then there is a $\delta > 0$ and an open interval I containing ξ such that $E^* = E \cap I$ is an infinite set and

$$|\pi(x)| \geq \delta > 0 \qquad \text{for } x \in I.$$

According to Theorem 5.3, there is an orthonormal sequence $\{\phi_n\}$ in $C(J)$ each of whose terms vanish outside I. Then

$$\|\pi\phi_m - \pi\phi_n\|^2 = \int_I \pi^2(x)[\phi_m(x) - \phi_n(x)]^2 \, d\psi(x)$$

$$\geq \delta^2 \|\phi_m - \phi_n\|^2,$$

hence

$$\|\pi\phi_m - \pi\phi_n\|^2 \geq 2\delta^2. \qquad (6.8)$$

Now write

$$\pi\phi_n \doteq \sum_{k=0}^{\infty} d_{nk} p_k$$

so that (cf. (6.5))

$$d_{nk} = \sum_{j=k-p}^{k+p} c_{nj} \pi_{jk}, \qquad c_{nj} = (\phi_n, p_j). \qquad (6.9)$$

By hypothesis (6.6), there is an integer K (independent of n) such that

$$|\pi_{nk}| \leqq \epsilon = \frac{\delta}{2(2p+1)} \qquad \text{for } k \geqq K.$$

With the aid of the preceding lemma, we conclude

$$\begin{aligned}
\|\pi\phi_m - \pi\phi_n\|^2 &\leqq \left\|\sum_{k=0}^{K-1}(d_{mk}-d_{nk})p_k\right\|^2 + \left\|\sum_{k=K}^{\infty}(d_{mk}-d_{nk})p_k\right\|^2 \\
&\leqq \sum_{k=0}^{K-1}(d_{mk}-d_{nk})^2 + 2\left(\sum_{k=K}^{\infty}d_{mk}^2 + \sum_{k=K}^{\infty}d_{nk}^2\right) \\
&\leqq \sum_{k=0}^{K-1}(d_{mk}-d_{nk})^2 + \delta^2.
\end{aligned}$$

But $\|\phi_n\|^2 = \sum_{k=0}^{\infty}c_{nk}^2 = 1$ so $\{c_{nk}\}_{n=1}^{\infty}$ is a bounded sequence for each k. By II-Theorem 2.1 (see II-Ex. 2.1), we know there is an increasing sequence of integers n_ν such that $\{c_{n_\nu k}\}_{\nu=1}^{\infty}$ converges for each k. Referring to (6.9), we see that $\{d_{n_\nu k}\}_{\nu=1}^{\infty}$ converges for each k. Therefore, finally, we conclude

$$\limsup_{\mu,\nu\to\infty}\left\|\pi\phi_{n_\mu} - \pi\phi_{n_\nu}\right\|^2 \leqq \delta^2$$

and this contradicts (6.8). ∎

If we choose $\pi(x) = (x-\sigma_1)\cdots(x-\sigma_p)$ where the σ_i are distinct real numbers, then (6.6) is necessary and sufficient for $E' \subset \{\sigma_1,\ldots,\sigma_p\}$ (E' denotes the derived set). Therefore we can restate Theorem 6.1 in the following form:

THEOREM 6.2   Let $\{c_n\}$ and $\{\lambda_n\}$ be bounded sequences. Then in order that $\mathfrak{S}(\psi)$ be a denumerable set whose derived set is $\{\sigma_1,\ldots,\sigma_p\}$, $\sigma_i \neq \sigma_j$ for $i \neq j$, it is necessary and sufficient that for every i,

$$\lim_{n\to\infty}\int_J \pi(x)p_n(x)p_{n+i}(x)\,d\psi(x) = 0$$

when $\pi(x) = (x-\sigma_1)\cdots(x-\sigma_p)$, and that this condition fail for at least one i if $\pi(x)$ is a non-zero polynomial of degree $< p$.

In order to relate Theorem 6.1 more explicitly with the coefficients in the recurrence formula, we observe that the recurrence formula for the orthonormal polynomials can be written (cf. Ex. 1.1).

$$xp_n(x) = \lambda_{n+1}^{1/2} p_{n-1}(x) + c_{n+1} p_n(x) + \lambda_{n+2}^{1/2} p_{n+1}(x), \qquad (n \geqq 0). \quad (6.10)$$

Taking $\pi(x) = x - \sigma$, we would have

$$\pi_{nk} = \begin{cases} \lambda_{n+2}^{1/2} & k = n + 1 \\ c_{n+1} - \sigma & k = n \\ \lambda_{n+1}^{1/2} & k = n - 1 \end{cases}$$

and, of course, $\pi_{nk} = 0$ for $|n - k| > 1$. Theorem 6.1 thus shows that $E' = \{\sigma\}$ if and only if

$$\lim_{n \to \infty} c_n = \sigma \quad \text{and} \quad \lim_{n \to \infty} \lambda_n = 0.$$

Thus Theorem 3.5 is a special case of Theorem 6.1.

We next introduce certain infinite matrices which will provide us with an alternative to (6.2).

First observe that (6.10) can be written

$$xp_n(x) = \sum_{k=0}^{\infty} \alpha_{nk} p_k(x) \qquad (6.11)$$

where

$$\alpha_{nk} = \alpha_{kn}, \quad \alpha_{nk} = 0 \qquad \text{if } |n - k| > 1$$

$$\alpha_{nn} = c_{n+1}, \qquad \alpha_{n,n+1} = \lambda_{n+2}^{1/2}, \qquad n \geqq 0.$$

With (6.11), we now associate the infinite matrix

$$\alpha = (\alpha_{ij})_{i,j=0}^{\infty} = \begin{bmatrix} c_1 & \lambda_2^{1/2} & 0 & 0 & 0 & \cdots \\ \lambda_2^{1/2} & c_2 & \lambda_3^{1/2} & 0 & 0 & \cdots \\ 0 & \lambda_3^{1/2} & c_3 & \lambda_4^{1/2} & 0 & \cdots \\ \cdot & \cdot & \cdot & \cdot & \cdot & \cdots \\ \cdot & \cdot & \cdot & \cdot & \cdot & \cdots \end{bmatrix} \qquad (6.12)$$

(An infinite matrix with this symmetric, tri-diagonal form is called a *Jacobi matrix*).

More generally, let $g \in C(J)$ and write

$$g(x)p_n(x) \doteq \sum_{k=0}^{\infty} g_{nk} p_k(x) \qquad (6.13)$$

so that $g_{nk} = (gp_n, p_k) = (p_n, gp_k) = g_{kn}$. We then consider the associated symmetric matrix

$$(g) = (g_{ij})_{i,j=0} = \begin{bmatrix} g_{00} & g_{01} & g_{02} & \cdots \\ g_{10} & g_{11} & g_{12} & \cdots \\ g_{20} & g_{21} & g_{22} & \cdots \\ \cdot & \cdot & \cdot & \cdots \\ \cdot & \cdot & \cdot & \cdots \end{bmatrix}$$

If also $f \in C(J)$ and $(f) = (f_{ij})$ is the associated matrix, we have

$$f(x)g(x)p_n(x) \doteq \sum_{k=0}^{\infty} h_{nk} p_k(x)$$

where

$$h_{nk} = (fgp_n, p_k) = (fp_n, gp_k).$$

According to Theorem 5.2

$$h_{nk} = \sum_{\nu=0}^{\infty} f_{n\nu} g_{\nu k}, \quad n, k = 0, 1, 2, \ldots. \tag{6.14}$$

This clearly suggests the formal matrix product, $(f_{ik})(g_{kj})$.

Therefore, with the obvious definitions of sums and products of infinite matrices (note that (6.14) always converges) we can write

$$(f + g) = (f) + (g), \qquad (fg) = (f)(g).$$

In particular, for any polynomial $\pi(x) = a_0 x^n + a_1 x^{n-1} + \cdots + a_n$,

$$(\pi) = a_0 \alpha^n + a_1 \alpha^{n-1} + \cdots + a_n I,$$

where $I = (\delta_{ij})$ is the identity matrix.
That is

$$(\pi_{ij})_{i,j=0}^{\infty} = (\pi) = \pi(\alpha). \tag{6.15}$$

For example, taking $\pi(x) = (x - a)(x - b)$, we obtain

$$(\pi) = \alpha^2 - (a + b)\alpha + abI$$
$$= \left(\sum_{k=0}^{\infty} \alpha_{ik} \alpha_{kj} - (a + b)\alpha_{ij} + ab\delta_{ij} \right)_{i,j=0}^{\infty}$$

so that

$$\pi_{nn} = \lambda_{n+1} + \lambda_{n+2} + (c_{n+1} - a)(c_{n+1} - b)$$

$$\pi_{n,n+1} = \lambda_{n+2}^{1/2}(c_{n+1} + c_{n+2} - a - b)$$

$$\pi_{n,n+2} = (\lambda_{n+2}\lambda_{n+3})^{1/2}.$$

We conclude this discussion of Krein's ideas by looking briefly at the connection between the preceding and the concept of completely continuous operators.

Let A denote the "multiplication operator" defined by

$$(Af)(x) = xf(x), \quad f \in C(J).$$

For any $g \in C(J)$, define the operator $g(A)$ by

$$[g(A)f](x) = g(x)f(x), \quad f \in C(J).$$

We deal henceforth with the topology induced on $C(J)$ by $\|\cdot\|$. Specifically, $\{\phi_n\}_{n=1}^{\infty}$ is *bounded* means there is an $M > 0$ such that $\|\phi_n\| \leq M$ ($n \geq 1$); $\{\phi_n\}_{n=1}^{\infty}$ is a *Cauchy sequence* means $\lim_{m,n\to\infty}\|\phi_m - \phi_n\| = 0$.

DEFINITION 6.1   A linear transformation $L: C(J) \to C(J)$ is called a *completely continuous operator* if for every bounded sequence $\{\phi_n\}$, $\{L\phi_n\}$ contains a Cauchy sub-sequence.

A re-examination of the proof of Theorem 6.1 reveals the truth of the following:

THEOREM 6.3   Let J be bounded and let $\pi(x)$ be a polynomial of degree p. Then (6.6) holds if and only if $\pi(A)$ is a completely continuous operator.

Proof It was shown in the proof of Theorem 6.1 that if (6.6) fails, then (6.7) follows and this shows that $\{\pi p_n\}$ contains no Cauchy subsequence.

Conversely suppose (6.6) is valid. Let $\{\phi_n\}$ be an arbitrary bounded sequence. We can assume without loss of generality that $\|\phi_n\| \leq 1$. The argument which begins with the line immediately following (6.8) can then be used virtually without change to show that $\{\pi(A)\phi_n\}$ contains a Cauchy subsequence. ∎

We leave as an exercise the task of locating appropriate arguments given (essentially) in the proof of Theorem 6.1 to prove our last theorem of Krein.

THEOREM **6.4** Let J be bounded and let $g \in C(J)$. Then $g(A)$ is completely continuous if and only if every limit point of E is a zero of g.

Exercises

6.1 Let $\lambda_n \to 0$. Prove that $E' = \{a,b\}$, $a \neq b$, if and only if $\{c_n\}$ is bounded and has two sequential limit points a and b.

6.2 Let $c_{2n} \to a$ and $c_{2n+1} \to b$. Prove that $E' = \{a,b\}$ if and only if $\lambda_n \to 0$.

6.3 (a) Find an example in which $E' = \{a,b\}$ and $\{c_n\}$ has two limit points, neither of them a or b.
(b) Find an example in which $E' = \{a,b\}$ and $\{c_n\}$ has four different limit points.

6.4 Let $\gamma_1, \gamma_2, \gamma_3, \ldots$ be distinct real numbers and let

$$\pi^{(k)}(x) = (x - \gamma_1) \cdots (x - \gamma_k).$$

Let $\pi_{mn}^{(k)}$ denote the corresponding coefficients given by (6.2). Derive the formula

$$\pi_{n,n+i}^{(k+1)} = a_{n-1}\pi_{n-1,n+i}^{(k)} + (b_n - \gamma_{k+1})\pi_{n,n+i}^{(k)} + a_n \pi_{n+1,n+i}^{(k)}.$$

6.5 Let $\lambda_n \to 0$ and $\{c_n\}$ be bounded. Use Ex. 6.4 to prove that $E' = \{\gamma_1, \gamma_2, \ldots, \gamma_p\}$ if and only if the set of sequential limit points of $\{c_n\}$ is $\{\gamma_1, \gamma_2, \ldots, \gamma_p\}$.

6.6 Show that if $\lambda_n \geq \lambda > 0$ for $k^2 \leq n \leq k(k+1)$ $(k = 1,2,3,\ldots)$, then E' is an infinite set.

6.7 Supply the formal proof for Theorem 6.4. (Note that the set of points t for which $g(t) \neq 0$ can be finite.)

6.8 Assume E is bounded. Prove directly (i.e. without recourse to Theorem 6.4) that if g is a continuous function without zeros in E, then $g(A)$ is not completely continuous.

6.9 Prove that if $\mathcal{L}$ is symmetric, then the weights of the Gauss quadrature formula satisfy $A_{n,n-k+1} = A_{nk}$.

CHAPTER V

Special Functions

1 General Remarks

In the remainder of this book, we will be concerned with the "special functions" aspect of orthogonal polynomials. In particular, we will discuss the properties of a large number of specific systems of orthogonal polynomials that have appeared in the literature.

Of necessity, we will no longer adhere to the controlled set of prerequisites used for the first four chapters but will make free use of standard results from classical analysis. No consistency will be attempted in the manner of presentation of the various special OPS. Only rarely will proofs be given. Occasionally a sketch or outline of a derivation may be indicated but frequently results will be stated accompanied by no more than a reference to the literature.

Whenever possible, standard notation will be used. When different notation is in common use for a particular OPS, we will usually follow the notation of Szegö [5] and the Bateman Manuscript Project, Erdélyi et. al. [1] (in that order of preference). The latter will be taken as the standard reference for special functions other than OPS.

We will say, "positive-definite OPS," to mean, "OPS with respect to a positive-definite moment functional." With a few exceptions, our discussion will deal only with positive-definite OPS.

Throughout the remainder of the book, the convention will be maintained that when a recurrence formula is written for an OPS $\{P_n(x)\}$ the initial conditions are $P_{-1}(x) = 0$, $P_0(x) = 1$, unless it is explicitly indicated otherwise.

2 The Classical Orthogonal Polynomials

The systems of orthogonal polynomials associated with the names of Hermite, Laguerre, and Jacobi (including the special cases named after

Tchebichef, Legendre, and Gegenbauer) are unquestionably the most extensively studied and widely applied systems. These three systems are called collectively the *classical orthogonal polynomials*.

The literature on these polynomials is enormous and we will present only the most basic facts concerning them. The most thorough single account of the classical polynomials is found in the treatise of Szegö [5]. The Bateman Manuscript Project, Erdélyi [1, vol. 2] contains a compact but detailed summary of their formal properties. Excellent accounts can be found in Jackson [1], Sansone [1], and Courant and Hilbert [1] (especially for the applications).

Because of the wide availability of properties of these polynomials, we will limit our discussion of them primarily to certain unifying properties which, while they are rather well known, are not generally found outside the original sources.

(A) *Definitions* The *Jacobi polynomials*, $P_n^{(\alpha,\beta)}(x)$, are the polynomials defined by the formula

$$P_n^{(\alpha,\beta)}(x) = (-2)^{-n}(n!)^{-1}(1-x)^{-\alpha}(1+x)^{-\beta}\frac{d^n}{dx^n}[(1-x)^{n+\alpha}(1+x)^{n+\beta}].$$

$$(2.1)$$

Here α and β are parameters which, for integrability purposes, are restricted to $\alpha > -1$, $\beta > -1$. However, many of the identities and other formal properties of these polynomials remain valid under the less restrictive condition that neither α nor β is a negative integer.

Among the many special cases that have received detailed attention, the following are the most important.

(i) The *Legendre* polynomials ($\alpha = \beta = 0$)

$$P_n(x) = P_n^{(0,0)}(x). \qquad (2.2)$$

(ii) The *Tchebichef* polynomials of the *first kind* ($\alpha = \beta = -1/2$)

$$T_n(x) = 2^{2n}\binom{2n}{n}^{-1}P_n^{(-\frac{1}{2},-\frac{1}{2})}(x). \qquad (2.3)$$

(iii) The *Tchebichef* polynomials of the *second kind* ($\alpha = \beta = -1/2$).

$$U_n(x) = 2^{2n}\binom{2n+1}{n+1}^{-1}P_n^{(\frac{1}{2},\frac{1}{2})}(x). \qquad (2.4)$$

(iv) The *Gegenbauer* (or *Ultraspherical*) polynomials ($\alpha = \beta$)

$$P_n^{(\lambda)}(x) = \binom{2\alpha}{\alpha}^{-1}\binom{n+2\alpha}{\alpha}P_n^{(\alpha,\alpha)}(x), \tag{2.5}$$

$$\alpha = \lambda - 1/2 \neq -1/2.$$

The formula (2.1) is usually called Rodrigues' formula in the case $\alpha = \beta = 0$ while the general case is called a Rodrigues' type formula. Legendre [1] investigated the polynomials which bear his name in 1785 while Rodrigues' formula appeared in 1816 (Rodrigues [1]). The general Jacobi polynomial was introduced by Jacobi [1] in 1859.

Using Leibniz' formula for the nth derivative of a product, we obtain from (2.1)

$$(-2)^n n! (1-x)^\alpha (1+x)^\beta P_n^{(\alpha,\beta)}(x)$$

$$= \sum_{k=0}^n \binom{n}{k} D^{n-k}(1-x)^{n+\alpha} D^k (1+x)^{n+\beta}$$

$$= (-1)^n (1-x)^\alpha (1+x)^\beta n! \sum_{k=0}^n \binom{n+\alpha}{n-k}\binom{n+\beta}{k}(x-1)^k(x+1)^{n-k}$$

(where we have written $D = d/dx$).
We thus obtain the explicit formula,

$$P_n^{(\alpha,\beta)}(x) = 2^{-n} \sum_{k=0}^n \binom{n+\alpha}{n-k}\binom{n+\beta}{k}(x-1)^k(x+1)^{n-k}. \tag{2.6}$$

We see that $P_n^{(\alpha,\beta)}(x)$ is a polynomial of degree n whose leading coefficient is

$$k_n = k_n(\alpha,\beta) = 2^{-n} \sum_{k=0}^n \binom{n+\alpha}{k}\binom{n+\beta}{n-k} = 2^{-n}\binom{2n+\alpha+\beta}{n}. \tag{2.7}$$

We also have

$$P_n^{(\alpha,\beta)}(-x) = (-1)^n P_n^{(\beta,\alpha)}(x), \tag{2.8}$$

$$P_n^{(\alpha,\beta)}(1) = \binom{n+\alpha}{n} \tag{2.9}$$

The *Laguerre polynomial*, $L_n^{(\alpha)}(x)$, can also be defined by a Rodrigues' type formula. Namely,

$$L_n^{(\alpha)}(x) = (n!)^{-1} x^{-\alpha} e^x \frac{d^n}{dx^n} [x^{n+\alpha} e^{-x}]. \qquad (2.10)$$

It is customary to require that $\alpha > -1$ but most formal relations remain valid if α is not a negative integer. The case $\alpha = 0$ is the one originally studied by Laguerre [1] although it occurred earlier in the works of Abel, Lagrange and Tchebichef. The notation,

$$L_n(x) = L_n^{(0)}(x),$$

is standard. The case of general α is due to Sonine [1] and $L_n^{(\alpha)}(x)$ is sometimes referred to as the *Sonine-Laguerre* or *generalized Laguerre* polynomial.

As with the Jacobi polynomials, an explicit formula for $L_n^{(\alpha)}(x)$ is obtained from (2.10) with the aid of Leibniz' formula. The result is

$$L_n^{(\alpha)}(x) = \sum_{k=0}^{n} \binom{n+\alpha}{n-k} \frac{(-x)^k}{k!}. \qquad (2.11)$$

We conclude that $L_n^{(\alpha)}(x)$ is a polynomial of degree n with leading coefficient

$$k_n = k_n(\alpha) = \frac{(-1)^n}{n!}. \qquad (2.12)$$

While the above definition is the most common, others are in use. For example, Courant-Hilbert [1] refers to $n! \, L_n(x)$ as the Laguerre polynomial (only the case $\alpha = 0$ is considered) while Jackson [1] calls $(-1)^n n! \, L_n^{(\alpha)}(x)$ the Laguerre polynomial.

The *Hermite polynomials*, $H_n(x)$, are defined by the Rodrigues' type formula

$$H_n(x) = (-1)^n e^{x^2} \frac{d^n}{dx^n} e^{-x^2}. \qquad (2.13)$$

Prior to Hermite's investigation [1], these polynomials were considered by Tchebichef [1] and first appeared in the famous *Mecanique Celeste* of Laplace.

Taylor's theorem yields the generating function (see I-Ex. 1.4)

$$e^{2xw-w^2} = \sum_{n=0}^{\infty} H_n(x) \frac{w^n}{n!}. \qquad (2.14)$$

Expanding e^{2xw} and e^{-w^2} as power series in w and taking the Cauchy product of the results yields

$$e^{2xw-w^2} = \sum_{n=0}^{\infty} \sum_{k=0}^{[n/2]} \frac{(-1)^k (2x)^{n-2k} w^n}{(n-2k)!\, k!}.$$

This provides the explicit formula

$$H_n(x) = n! \sum_{k=0}^{[n/2]} \frac{(-1)^k (2x)^{n-2k}}{(n-2k)!\, k!}. \tag{2.15}$$

In these formulas, $[x]$ denotes the largest integer not exceeding x. $H_n(x)$ is thus seen to be a polynomial of degree n whose leading coefficient is

$$k_n = 2^n. \tag{2.16}$$

The most common alternative to the above terminology is to replace $\exp(-x^2)$ by $\exp(-x^2/2)$ in (2.13). This yields a polynomial, $He_n(x)$, which can be expressed in terms of $H_n(x)$ by

$$He_n(x) = 2^{-n/2} H_n(2^{-1/2}x).$$

$He_n(x)$ is called the Hermite polynomial for example by Jackson [1] and seems to be the preferred form for applications to statistics (see Cramér [1]).

(B) *Orthogonality*   Note that in each of the three cases, we have defined a sequence of polynomials by a formula of the type

$$P_n(x) = K_n^{-1}[w(x)]^{-1} D^n[\rho^n(x)w(x)], \qquad n = 0, 1, 2, \ldots, \tag{2.17}$$

where for the Jacobi, Laguerre, and Hermite cases, respectively:
   (i) $K_n = (-2)^n n!$, $n!$, and $(-1)^n$;
   (ii) $\rho(x)$ is a polynomial independent of n and of degree 2, 1, and 0;
   (iii) $w(x)$ is positive and integrable over (a,b) where (a,b) is $(-1,1)$, $(0,\infty)$, and $(-\infty,\infty)$, respectively;
   (iv) $D^k[\rho^n(x)w(x)]$ vanishes for $x = a$ and $x = b$, $0 \leq k < n$.
(Here, and in (iii), the conditions, $\alpha > -1$, $\beta > -1$ in the Jacobi case and $\alpha > -1$ in the Laguerre case must be imposed.)
   For nonnegative integers m and n, write

$$I_{mn} = \int_a^b x^m P_n(x)w(x)\, dx = K_n^{-1} \int_a^b x^m D^n[\rho^n(x)w(x)]\, dx.$$

Integrating by parts and using (iv), we find

$$K_n I_{mn} = x^m D^{n-1} [\rho^n(x)w(x)]\big|_a^b - m \int_a^b x^{m-1} D^{n-1} [\rho^n(x)w(x)]\, dx$$

$$= -m \int_a^b x^{m-1} D^{n-1} [\rho^n(x)w(x)]\, dx.$$

We assume that $0 \leqq m \leqq n$. If the above procedure is repeated, we then obtain after m such steps,

$$K_n I_{mn} = (-1)^m m! \int_a^b D^{n-m} [\rho^n(x)w(x)]\, dx.$$

Then if $m < n$, one more integration yields

$$K_n I_{mn} = (-1)^m m! \, D^{n-m-1} [\rho^n(x)w(x)]\big|_a^b = 0.$$

On the other hand, if $m = n$, we have

$$K_n I_{nn} = (-1)^n n! \int_a^b \rho^n(x)w(x)\, dx.$$

Thus in the three cases, we have specifically,

(Jacobi) $\quad I_{nn} = 2^{-n} \int_{-1}^{1} (1 - x)^{n+\alpha}(1 + x)^{n+\beta}\, dx$

$$= 2^{n+\alpha+\beta+1} B(n + \alpha + 1, n + \beta + 1),$$

where B denotes the *beta function* which can be expressed in terms of the gamma function by

$$B(x,y) = \frac{\Gamma(x)\Gamma(y)}{\Gamma(x + y)};$$

(Laguerre) $\quad I_{nn} = (-1)^n \int_0^{\infty} x^{n+\alpha} e^{-x}\, dx = (-1)^n \Gamma(n + \alpha + 1);$

(Hermite) $\quad I_{nn} = n! \int_{-\infty}^{\infty} e^{-x^2}\, dx = n! \sqrt{\pi}.$

Referring to (2.7), (2.12), and (2.16) for the leading coefficients of each of the three polynomials, we then can write the explicit orthogonality relations,

$$\int_{-1}^{1} P_m^{(\alpha,\beta)}(x) P_n^{(\alpha,\beta)}(x)(1-x)^\alpha (1+x)^\beta \, dx$$

$$= \frac{2^{\alpha+\beta+1}\Gamma(n+\alpha+1)\Gamma(n+\beta+1)}{(2n+\alpha+\beta+1)\Gamma(n+\alpha+\beta+1)n!}\delta_{mn},$$

$$(\alpha > -1, \beta > -1); \qquad (2.18)$$

$$\int_0^\infty L_m^{(\alpha)}(x) L_n^{(\alpha)}(x) x^\alpha e^{-x} \, dx = \frac{\Gamma(n+\alpha+1)}{n!}\delta_{mn}, \qquad \alpha > -1;$$

$$\int_{-\infty}^\infty H_m(x) H_n(x) e^{-x^2} \, dx = 2^n n! \sqrt{\pi}\, \delta_{mn}$$

(C) *Derivatives and differential equations*   As is suggested by the proof of the orthogonality relations, a large number of properties are shared by the three classical systems and many of these can be derived using the Rodrigues' type formula, (2.17).

Thus $y = P_n(x)$ satisfies the self-adjoint differential equation

$$\frac{d}{dx}[\rho(x)w(x)\frac{dy}{dx}] - \lambda_n w(x)y = 0, \qquad n = 0, 1, 2, \ldots, \qquad (2.19)$$

where the eigenvalues, λ_n, are given by

$$\lambda_n = -n(n+\alpha+\beta+1), \qquad y = P_n^{(\alpha,\beta)}(x),$$

$$\lambda_n = -n, \qquad\qquad\qquad y = L_n^{(\alpha)}(x),$$

$$\lambda_n = -2n, \qquad\qquad\qquad y = H_n(x).$$

The derivation of (2.19) based upon (2.17) is due to Tricomi (a sketch of the proof is given in Erdélyi [1, vol. 2]), however the three cases have long been known individually. The differential equation is the major reason for the importance of the classical polynomials for applications to boundary value problems in classical physics and quantum mechanics. For example, see Churchill [1], Courant-Hilbert [1], Jackson [1].

Written out explicitly, the three cases can be put in the forms:

$$(1 - x^2)y'' + [\beta - \alpha - (\alpha + \beta + 2)x]y' + n(n + \alpha + \beta + 1)y = 0,$$

$$y = P_n^{(\alpha,\beta)}(x)$$

$$(2.20)$$

$$xy'' + (\alpha + 1 - x)y' + ny = 0, \qquad y = L_n^{(\alpha)}(x)$$

$$y'' - 2xy' + 2ny = 0, \qquad y = H_n(x)$$

Tricomi [1] has also used (2.17) to derive the differentiation formula

$$\rho(x)\frac{dP_n(x)}{dx} = [A_n + \frac{1}{2}nx\rho''(x)]P_n(x) + B_n P_{n-1}(x),$$

where A_n and B_n are certain constants (see Erdélyi [1, vol. 2]). The specific formulas for the three cases (again previously known individually) are

$$(2n + \alpha + \beta)(1 - x^2)\frac{dP_n^{(\alpha,\beta)}(x)}{dx} = n[\alpha - \beta - (2n + \alpha + \beta)x]P_n^{(\alpha,\beta)}(x)$$

$$+ 2(n + \alpha)(n + \beta)P_{n-1}^{(\alpha,\beta)}(x)$$

$$x\frac{dL_n^{(\alpha)}(x)}{dx} = nL_n^{(\alpha)}(x) - (n + \alpha)L_{n-1}^{(\alpha)}(x)$$

$$(2.21)$$

$$\frac{dH_n(x)}{dx} = 2nH_{n-1}(x)$$

For the Laguerre and Jacobi polynomials, there are simpler formulas available for the derivatives. The formula

$$\frac{dL_n^{(\alpha)}(x)}{dx} = -L_{n-1}^{(\alpha+1)}(x) \qquad (2.22)$$

can be verified from (2.11).

The derivation of a corresponding formula for the Jacobi polynomials is more troublesome. The usual method is to observe that the differential equation for the Jacobi polynomials is one of Gauss' hypergeometric type. Thus $P_n^{(\alpha,\beta)}(x)$ can be expressed as a hypergeometric function. This hypergeometric representation can then be used to verify the formula

$$\frac{dP_n^{(\alpha,\beta)}(x)}{dx} = \frac{1}{2}(n + \alpha + \beta + 1)P_{n-1}^{(\alpha+1,\beta+1)}(x). \qquad (2.23)$$

(see Szegö [5].)

We note that (2.22), (2.23) and the last equation of (2.21) show that in all three cases, the sequence of derivatives forms another OPS.

(*D*) *Characterizations of the classical orthogonal polynomials*   Many of the previous properties which are common to the three systems characterize these polynomials in the sense that the only positive-definite orthogonal polynomials that have these properties are the classical orthogonal polynomials or are reducible to them by a linear change of independent variable.

The first such characterization was obtained by S. Bochner [1]. The differential equation (2.19) can be written in the form

$$a_2(x)y'' + a_1(x)y' + a_0(x)y + \lambda y = 0, \tag{2.24}$$

where $a_0(x)$, $a_1(x)$, $a_2(x)$ are polynomials of degrees at most 0, 1, and 2, respectively, and $\lambda \neq 0$.

Bochner began with (2.24) where $a_0(x)$, $a_1(x)$, and $a_2(x)$ are polynomials, but without any assumptions concerning their degrees, and determined all cases such that for each integer $n \geq 0$, there is an eigenvalue $\lambda = \lambda_n$ for which there is a corresponding solution which is a polynomial of degree n.

If for some $\lambda = \lambda_0 \neq 0$, (2.24) has a nontrivial constant solution, it is easy to see that $a_0(x)$ must be a constant which, without loss of generality, can be taken to be 0. If for some $\lambda = \lambda_1 \neq 0$, (2.24) has a linear polynomial solution, then it is again immediately seen that $a_1(x)$ must itself be a linear polynomial. Finally, it is similarly deduced that $a_2(x)$ must be a quadratic polynomial.

Now by a detailed analysis of the possible cases, Bochner showed that the only possibilities are those for which the eigensolutions can be reduced by a linear change of variable (apart from constant factors) to

(a) $y = P_n^{(\alpha,\beta)}(x)$ with the restrictions that α, β, and $\alpha + \beta + 1$ are not negative integers;

(b) $y = L_n^{(\alpha)}(x)$ with the restriction, $\alpha \neq -1, -2, -3, \ldots$;

(c) $y = H_n(x)$;

(d) $y = x^n$ (there is a case in which one can have $y = x^n + b_m x^m$ for finitely many values of n);

(e) $y = \sum_{k=0}^n \binom{n}{k}(-1)^k (\alpha + n)_k x^k$, where $\alpha \neq -1, -2, -3, \ldots$, and we have written, $(a)_0 = 1$, $(a)_n = a(a + 1) \cdots (a + n - 1)$ for $n \geq 1$.

It is clear that the polynomials of case (d) cannot form an OPS. In case (e), it is readily verified that the polynomial for $n = 2$ has nonreal zeros. Thus the only positive-definite orthogonal polynomials that arise as eigensolutions of (2.24) are the classical polynomials.

Although the polynomials of case (e) are not orthogonal in the positive-definite sense, they are orthogonal in the quasi-definite sense. This was apparently first observed by H. L. Krall [3]. These polynomials are called *Bessel polynomials* and are discussed in Chapter VI.

It was noted earlier that the derivatives of the classical orthogonal polynomials are again orthogonal polynomials. It was first proved by W. Hahn [1] that this property also characterizes the classical polynomials. More precisely, Hahn showed that if $\{P_n(x)\}$ is a positive-definite OPS such that the sequence of derivatives, $\{P'_n(x)\}$, is also a positive-definite OPS, then $\{P_n(x)\}$ can be reduced to one of the three classical systems by a linear change of the variable x.

Hahn's method was to begin with a pair of recurrence relations,

$$P_n(x) = (x - c_n)P_{n-1}(x) - \lambda_n P_{n-2}(x),$$

$$\frac{P'_n(x)}{n} = (x - d_n)\frac{P'_{n-1}(x)}{n-1} - \nu_n \frac{P''_{n-2}(x)}{n-2}.$$

By differentiation of the first of these, $P'_n(x)$ can be eliminated from the second. Then by a judicious repetition of such steps, Hahn was able to obtain eventually certain recurrence relations from which it could be reasoned that $P_n(x)$ satisfied a differential equation of the form (2.24). (The actual procedure is rather complex and the various intermediate equations are very complicated.) Hahn then essentially duplicated Bochner's work to conclude that $\{P_n(x)\}$ must be one of the three classical OPS.

It was later observed by Krall [3] and F. S. Beale [1] that Hahn's procedure applies equally well to the general quasi-definite case. Thus the only OPS whose derivatives form an OPS in the quasi-definite sense are the classical OPS and the Bessel polynomials of the preceding case (e). Hahn had already noted that the latter were not orthogonal in the positive-definite sense.

Hahn later [2] extended his results by proving that if $\{P_n(x)\}$ is an OPS such that $\{P_n^{(r)}(x)\}$ is also an OPS for some fixed integer $r \geqq 1$, then $\{P_n(x)\}$ must be one of the classical OPS. Earlier, Krall [1] had shown that if $\{P_n^{(r)}(x)\}$ is an OPS and $\{P_n(x)\}$ is an OPS over a *bounded* interval, then $\{P_n(x)\}$ was essentially the Jacobi OPS.

A third characterization of the classical orthogonal polynomials was suggested by Tricomi [1] and a complete proof was supplied by Ebert [1] and by Cryer [1] who pointed out and corrected an oversight in Tricomi's analysis. This characterization is that the only polynomial sequences that have Rodrigues' type formulas (2.17) are essentially the Hermite, Laguerre, Jacobi and Bessel polynomials with the usual restrictions on the parameters appearing in the latter three relaxed. In particular then, the only positive-definite OPS are the classical ones.

The method of proof is to consider (2.17) under the assumption that $\rho(x)$ is a polynomial and $P_n(x)$ is a polynomial of degree n. Examination of the cases $n = 1$ and $n = 2$ then shows that the degree of $\rho(x)$ cannot exceed 2 and that w must satisfy a differential equation of "Pearson type":

$$\frac{w'(x)}{w(x)} = \frac{ax + b}{\rho(x)}. \tag{2.25}$$

The solution of (2.25) is elementary and it is found that when $\rho(x)$ is constant, linear, or a quadratic with real, distinct zeros, the solution of (2.25) can be transformed by a linear change of variable into the forms corresponding to the Hermite, Laguerre and Jacobi polynomials, respectively.

However, as Cryer shows, if $\rho(x)$ has non-real zeros, (2.25) leads to Jacobi polynomials with nonreal parameters while if $\rho(x)$ has repeated zeros, it leads to Bessel polynomials.

Actually, Hildebrandt [1] had partially anticipated the above in his study of (2.25) from the viewpoint of the Gram-Charlier series in statistics. Hildebrandt shows that if w is a non-trivial solution of (2.25), then (2.17) generates a sequence of polynomials, $P_n(x)$, of degree at most n. He thus generalizes the work of Romanovsky [1], [2] who had studied a number of special cases and obtained the Rodrigues' type formula for not only the classical polynomials but the Bessel polynomials as well.

Hildebrandt does not investigate the question of when $P_n(x)$ is exactly of degree n nor the question of orthogonality in the general case. However he does obtain a number of formulas for the general polynomials including the three term formula! (Note that this was 1931 and Favard's paper on the three term recurrence formula did not appear until 1935.)

A fourth characterization was suggested in a conjecture of Karlin and Szegö [1]. Al-Salam and Chihara [1] have shown that the classical polynomials are characterized by orthogonality and the existence of a differentiation formula of the form

$$\pi(x) P'_n(x) = (\alpha_n x + \beta_n) P_n(x) + \gamma_n P_{n-1}(x)$$

where $\pi(x)$ is a fixed polynomial (cf. (2.21)). This result can be used to verify the conjecture of Karlin and Szegö whose paper should be consulted for a conjecture on yet another possible characterization that remains an open question.

(E) *Recurrence relations* By virtue of their orthogonality, the classical polynomials satisfy three term recurrence relations of the form I-(4.2):

$$P_{n+1}(x) = (A_n x + B_n) P_n(x) - C_n P_{n-1}(x), \qquad n \geqq 0,$$

$$P_{-1}(x) = 0, \qquad P_0(x) = 1.$$

The coefficients can be computed using I-(4.3). This yields A_n quickly in all three cases. However, for the Jacobi polynomials, B_n and C_n can be computed more conveniently by setting $x = -1$ and $x = 1$ and using (2.8) and (2.9). The somewhat unwieldy result can be written

$$2n(n + \alpha + \beta)(2n + \alpha + \beta - 2) P_n^{(\alpha, \beta)}(x) = (2n + \alpha + \beta - 1)$$

$$\cdot [(2n + \alpha + \beta)(2n + \alpha + \beta - 2)x + \alpha^2 - \beta^2] P_{n-1}^{(\alpha, \beta)}(x)$$

$$-2(n + \alpha - 1)(n + \beta - 1)(2n + \alpha + \beta) P_{n-2}^{(\alpha, \beta)}(x), \qquad n \geqq 1. \quad (2.26)$$

For the Laguerre polynomials, B_n and C_n can be computed quickly by using (2.11) and comparing coefficients of x^n and x^{n-1} in the recurrence. This yields

$$nL_n^{(\alpha)}(x) = (2n + \alpha - 1 - x)L_{n-1}^{(\alpha)}(x) - (n + \alpha - 1)L_{n-2}^{(\alpha)}(x), \qquad n \geqq 1. \quad (2.27)$$

In the case of the Hermite polynomials, we have $H_n(-x) = (-1)^n H_n(x)$ so that $B_n = 0$. C_n can be computed by comparison of coefficients of x^{n-1}. Alternately, one can verify that $C_n = A_n I_{nn}/I_{n-1,n-1} = 2n$. Thus

$$H_{n+1}(x) = 2xH_n(x) - 2nH_{n-1}(x), \qquad n \geqq 0. \quad (2.28)$$

For the corresponding *monic* polynomials, $\hat{P}_n(x) = k_n^{-1} P_n(x)$, these recurrences take the forms

$$\hat{P}_n^{(\alpha, \beta)}(x) = (x - c_n) \hat{P}_{n-1}^{(\alpha, \beta)}(x) - \lambda_n \hat{P}_{n-2}^{(\alpha, \beta)}(x), \qquad (2.29)$$

where

$$c_{n+1} = \frac{\beta^2 - \alpha^2}{(2n + \alpha + \beta)(2n + \alpha + \beta + 2)}$$

$$\lambda_{n+1} = \frac{4n(n + \alpha)(n + \beta)(n + \alpha + \beta)}{(2n + \alpha + \beta)^2 (2n + \alpha + \beta + 1)(2n + \alpha + \beta - 1)}$$

except that when $\alpha = -\beta$, $c_1 = (\beta - \alpha)/(\alpha + \beta + 2)$.

$$\hat{L}_n^{(\alpha)}(x) = (x - 2n - \alpha + 1)\hat{L}_{n-1}^{(\alpha)}(x) - (n - 1)(n + \alpha - 1)\hat{L}_{n-2}^{(\alpha)}(x), \quad (2.30)$$

$$\hat{H}_n(x) = x\hat{H}_{n-1}(x) - \frac{1}{2}(n - 1)\hat{H}_{n-2}(x). \quad (2.31)$$

(*F*) *Generating functions* There are a large number of generating functions known for the classical polynomials. By "generating function" for $\{P_n(x)\}$, we mean here a function F of two variables that has a formal Taylor's expansion of the form

$$F(x, w) \sim \sum_{n=0}^{\infty} a_n P_n(x)w^n,$$

where $\{a_n\}$ is a known sequence of constants.

The classical generating function for the Jacobi polynomials is

$$2^{\alpha+\beta}R^{-1}(1 - w + R)^{-\alpha}(1 + w + R)^{-\beta} = \sum_{n=0}^{\infty} P_n^{(\alpha,\beta)}(x)w^n, \quad (2.32)$$

where

$$R = (1 - 2xw + w^2)^{1/2}.$$

This generating function was obtained originally by Jacobi. For references and several different proofs, see Szegö [5]. For an elementary proof, see Carlitz [9]

For $\alpha = \beta$, (2.32) yields a generating function for the Gegenbauer polynomials. However, in this case there is the simpler generating function due to Gegenbauer:

$$(1 - 2xw + w^2)^{-\lambda} = \sum_{n=0}^{\infty} P_n^{(\lambda)}(x)w^n. \quad (2.33)$$

In turn, (2.33) yields for $\lambda = 1/2$ and 1, respectively, generating functions for the Legendre polynomials and Tchebichef polynomials of the second kind:

$$(1 - 2xw + w^2)^{-1/2} = \sum_{n=0}^{\infty} P_n(x)w^n, \quad (2.34)$$

$$(1 - 2xw + w^2)^{-1} = \sum_{n=0}^{\infty} U_n(x)w^n. \quad (2.35)$$

Using the readily verified limit

$$\lim_{\lambda \to 0} \lambda^{-1} P_n^{(\lambda)}(x) = \frac{2}{n} T_n(x),$$

one can also obtain from (2.33) a generating function for the Tchebichef polynomials of the first kind:

$$\log(1 - 2xw + w^2)^{-1} = 2 \sum_{n=1}^{\infty} n^{-1} T_n(x) w^n. \tag{2.36}$$

However, there is the simpler algebraic generating function (see I-Ex. 4.11)

$$\frac{1 - xw}{1 - 2xw + w^2} = \sum_{n=0}^{\infty} T_n(x) w^n. \tag{2.37}$$

The most common generating function for the Laguerre polynomials is

$$(1 - w)^{-\alpha-1} \exp \frac{-xw}{1 - w} = \sum_{n=0}^{\infty} L_n^{(\alpha)}(x) w^n. \tag{2.38}$$

This may be proved by expanding the left side as a series in w and using (2.11) to identify the coefficients in the resulting expansion. Alternately, one can use the orthogonality property (cf. I-Ex. 8.5).

For the Hermite polynomials, the standard generating function is

$$e^{2xw - w^2} = \sum_{n=0}^{\infty} \frac{H_n(x)}{n!} w^n, \tag{2.39}$$

which follows directly from Taylor's theorem and (2.13).

For further examples of generating functions of varying degrees of utility, see Erdélyi [1, vol. 3] and Rainville [1]. For applications, see also Boas and Buck [1].

(G) *Some close relatives of the classical polynomials*   Since

$$(1 - x)^{\alpha}(1 + x)^{\beta+1} = (1 + x)(1 - x)^{\alpha}(1 + x)^{\beta}, \qquad -1 < x < 1,$$

it follows that $\{P_n^{(\alpha,\beta+1)}(x)\}$ is a sequence of kernel polynomials corresponding to $\{P_n^{(\alpha,\beta)}(x)\}$ with K-parameter $\kappa = -1$. Taking due account of the leading coefficients, I-(7.3) can be used to express $P_n^{(\alpha,\beta+1)}(x)$ in terms of $P_n^{(\alpha,\beta)}(x)$ and $P_{n+1}^{(\alpha,\beta)}(x)$.

Making the change of variable $x = 2u - 1$ ($0 \leq u \leq 1$), we thus find that $\{P_n^{(\alpha,\beta)}(2u - 1)\}$ and $\{P_n^{(\alpha,\beta+1)}(2u - 1)\}$ are OPS with respect to the weight functions, $u^\beta(1-u)^\alpha$ and $u^{\beta+1}(1-u)^\alpha$, respectively. Referring to I-§8, we then set

$$S_{2m}(x) = P_m^{(\alpha,\beta)}(2x^2-1), \qquad S_{2m+1}(x) = xP_m^{(\alpha,\beta+1)}(2x^2-1) \quad (2.40)$$

and conclude that $\{S_n(x)\}$ is a symmetric OPS with respect to the weight function

$$|x|^{2\beta+1}(1-x^2)^\alpha, \qquad -1 < x < 1.$$

The case $\alpha = 0$ is due to Szegö [1] while the case $\beta = -1/2$ is a well known result concerning Gegenbauer polynomials.

The recurrence formula for the corresponding monic polynomials is

$$\hat{S}_n(x) = x\hat{S}_{n-1}(x) - \gamma_n \hat{S}_{n-2}(x), \qquad n \geq 1,$$

$$\gamma_{2m} = \frac{(m+\beta)(m+\alpha+\beta)}{(2m+\alpha+\beta-1)(2m+\alpha+\beta)}, \qquad (2.41)$$

$$\gamma_{2m+1} = \frac{m(m+\alpha)}{(2m+\alpha+\beta)(2m+\alpha+\beta+1)},$$

and can be verified with the aid of I-Theorem 9.1.

Asymptotic properties of these polynomials are included in the studies of Konoplev [1], [2]. For tables for the Gauss quadrature formulas corresponding to the case $\alpha = 0$, see Stroud and Secrest [2].

For a generalization of (2.40) to polynomials $R_n(x)$ satisfying the recurrence formula

$$R_n(x) = [x + (-1)^n c]R_{n-1}(x) - \gamma_n R_{n-2}(x), \qquad (2.42)$$

where γ_n is given by (2.41), see Chihara [4]. The weight function is

$$w(x) = |x + \alpha_0||x^2 - \alpha_0^2|^\beta(1 + \alpha_0^2 - x^2), \qquad \alpha_0^2 \leq x^2 \leq 1 + \alpha_0^2,$$

with $w(x) = 0$ otherwise.

The observation that $\{L_n^{(\alpha+1)}(x)\}$ is a kernel OPS corresponding to $\{L_n^{(\alpha)}(x)\}$ with K-parameter $\kappa = 0$ leads to defining

$$H_{2n}^{(\mu)}(x) = (-1)^n 2^{2n} n!\, L_n^{(\mu-1/2)}(x^2)$$
$$H_{2n+1}^{(\mu)}(x) = (-1)^n 2^{2n+1} n!\, L_n^{(\mu+1/2)}(x^2) \qquad \mu > -\frac{1}{2}. \quad (2.43)$$

The leading coefficient of $H_n^{(\mu)}(x)$ is 2^n. $\{H_n^{(\mu)}(x)\}$ is an OPS with respect to the weight function

$$|x|^{2\mu}e^{-x^2}, \qquad -\infty < x < \infty.$$

The $H_n^{(\mu)}(x)$ are called *generalized Hermite polynomials* since we have

$$H_n^{(0)}(x) = H_n(x).$$

These polynomials were introduced by Szegö [5, Problem 25] who also gave the differential equation

$$xy'' + 2(\mu - x^2)y' + (2nx - \theta_n x^{-1})y = 0, \qquad y = H_n^{(\mu)}(x), \quad (2.44)$$

where $\theta_{2m} = 0,\ \theta_{2m+1} = 2\mu$.

The explicit orthogonality relation is

$$\int_{-\infty}^{\infty} H_m^{(\mu)}(x)H_n^{(\mu)}(x)|x|^{2\mu}e^{-x^2}\,dx = 2^{2n}\left[\frac{n}{2}\right]!\,\Gamma\left(\left[\frac{n+1}{2}\right] + \mu + \frac{1}{2}\right)\delta_{mn},$$
$$(2.45)$$

where $[x]$ denotes the greatest integer function. The recurrence formula

$$H_{n+1}^{(\mu)}(x) = 2xH_n^{(\mu)}(x) - 2(n + \theta_n)H_{n-1}^{(\mu)}(x), \qquad n \geqq 0, \quad (2.46)$$

is readily verified, as is the differentiation formula

$$\frac{d}{dx}H_n^{(\mu)}(x) = 2nH_{n-1}^{(\mu)}(x) + 2(n-1)\theta_n x^{-1}H_{n-2}^{(\mu)}(x). \quad (2.47)$$

Many other well known formulas for the Hermite polynomials have analogues for the generalized case (Chihara [1]). For example, there is a sort of Rodrigues' type formula

$$H_n^{(\mu)}(x) = x^{-2\mu}e^{x^2}\frac{d^n}{dx^n}[x^{n+2\mu}e^{-x^2}K_n^{(\mu)}(x)], \quad (2.48)$$

where

$$K_{2m}^{(\mu)}(x) = \frac{(-1)^m}{(\mu+1)_m}\,{}_1F_1(m, m + \mu + 1, x^2)$$

$$K_{2m+1}^{(\mu)}(x) = \frac{(-1)^m}{(\mu+1)_{m+1}}x\,{}_1F_1(m + 1, m + \mu + 2; x^2).$$

This can be verified with the aid of Kummer's transformation,

$$e^{-x} {}_1F_1(a, c; x) = {}_1F_1(c - a, c; -x),$$

by using (2.43) to express $H_n^{(\mu)}(x)$ in terms of confluent hypergeometric functions (see Erdélyi [1, (10.12.14)]).

The generating function

$$(1 + 2xw + 4w^2)(1 + 4w^2)^{-(\mu+3/2)} \exp\{4x^2 w^2 (1 + 4w^2)^{-1}\}$$
$$= \sum_{n=0}^{\infty} H_n^{(\mu)}(x) \frac{w^n}{[n/2]!} \tag{2.49}$$

can be derived from (2.38) and (2.43).

The generalized Hermite polynomials have been mentioned in connection with Gauss quadrature formulas by Shao, Chen and Frank [1] and Stroud and Secrest [1], [2].

(*H*) *Miscellaneous remarks and references*   The Hamburger moment problems associated with the classical orthogonal polynomials and their relatives in (G) are all determined. For the Jacobi polynomials, this is an immediate consequence of the boundedness of the spectrum. In the Laguerre and Hermite cases, this can be deduced from the moments (see II-Ex. 6.3) or from the recurrence formula (see VI-(1.14)). A theorem of M. Riesz [1] then asserts that these polynomials form complete orthogonal systems in the appropriate L_2 spaces. Direct proofs of the completeness of the Laguerre and Hermite polynomials can be found in Szegö [5]. The completeness also follows from a theorem of Hewitt [1] which can also be applied to the generalized Hermite polynomials.

Various questions regarding expansions in series of classical orthogonal polynomials are discussed in Alexits [1], Erdélyi [1, vol. 2], Sansone [1], and Szegö [5]. The general literature on this subject is far too extensive to give even a partial representative listing, but a rather complete guide to papers published up to 1939 is Shohat, Hille and Tamarkin [1]. Some examples of recent investigations appear in Haimo [1]. A few other samples of recent work are Askey [2], [4], Askey and Wainger [1], [2], Pollard [1], Muckenhoupt [1], [2], [3], Muckenhoupt and Stein [1].

Applications of the classical orthogonal polynomials to boundary value problems in classical physics are discussed in Churchill [1], Courant-Hilbert [1], Hobson [1], Indritz [1], Kellog [1], Jackson [1], Whittaker and Watson [1]. For some recent investigations and further references, see Haimo [1]. Applications of the Hermite and Laguerre polynomials to quantum theory can be found in Indritz [1], Jackson [1], Weyl [1], and of course texts in physics.

For applications of the Hermite polynomials in statistics, see Cramér [1], Hildebrand [1], Feller [1]. Karlin and McGregor [3], [8] discuss the application of Jacobi and Laguerre polynomials in the study of certain Markov processes. Morrison [2] considers applications of the Laguerre polynomials to problems in electrical engineering. Leipnik [1] mentions Hermite and Laguerre polynomials in connection with noise problems. For applications of Jacobi polynomials with negative parameters, see Coulton [1].

Many texts on approximation theory give good accounts of the role played by the classical polynomials. See Cheney [1], Davis [1], Hildebrand [1], Natanson [1]. For Gauss quadratures in particular, see Krylov [1] and Stroud and Secrest [2]. Snyder [1] discusses the Tchebichef polynomials in detail.

A variety of problems of current interest dealing with the classical orthogonal polynomials are treated by Askey [9].

3 The Hahn Class of Orthogonal Polynomials

W. Hahn [4] generalized the classical orthogonal polynomials by generalizing the properties that characterize them. Specifically, let q and ω be fixed numbers with $q \neq 1$ and define a linear operator $\mathbf{L}$ by

$$\mathbf{L}f(x) = \frac{f(qx + \omega) - f(x)}{(q - 1)x + \omega}.$$

Note that the ordinary derivative operator is the limiting case $\omega = 0$ and $q \to 1$ of $\mathbf{L}$.

Hahn now poses the following problems. To determine all OPS $\{P_n(x)\}$ such that:

(A) $\{\mathbf{L}P_n(x)\}$ is also an OPS;

(B) $\{P_n(x)\}$ satisfies an operator equation of the form,

$$(\alpha x^2 + \beta x + \gamma)\mathbf{L}^2 P_n(x) + (\delta x + \epsilon)\mathbf{L}P_n(x) + \zeta P_n(x) = 0,$$

where α, β, γ, δ, ϵ, and ζ are constants;

(C) $P_n(x) = [w(x)]^{-1}\mathbf{L}^n[X_0(x)X_1(x)\cdots X_n(x)w(x)]$, where
$X_0(x)$ is a polynomials independent of n,
$X_i(x) = X_{i+1}(qx + \omega)$,
$w(x)$ is independent of n;

(D) $P_n(x) = \sum_{i=0}^{n} a_{ni}x^i$, where

$$\frac{a_{ni}}{a_{n,i-1}} = Q\left(\frac{q^n - 1}{q - 1}, \frac{q^i - 1}{q - 1}\right),$$

$Q(x, y)$ being a rational function of x and y;

(E) The moment sequence $\{\mu_n\}$ with respect to which $\{P_n(x)\}$ is an OPS satisfies a recurrence of the form

$$\mu_n = \frac{a + bq^n}{c + dq^n}\mu_{n-1}, \qquad ad - bc \neq 0.$$

Hahn proved that all five conditions determine one and the same class of orthogonal polynomials and that these polynomials may be constructed in terms of basic hypergeometric series (see Erdélyi [1, vol. 1], Hahn [6]). Since the derivative is a limiting case of Hahn's operator, the classical orthogonal polynomials are included in Hahn's class as limiting cases (note in particular, conditions (A), (B), and (C)).

Hahn's proofs are too involved to present here but a number of special cases studied individually in the literature are included among the Hahn class and these will be noted as they are discussed.

Hahn's operator includes as a special case, $q = 1$ and $\omega = 1$, the finite difference operator

$$\Delta f(x) = f(x + 1) - f(x).$$

For this special case, Erdélyi and Weber [1] obtained explicitly the corresponding orthogonal polynomials by working with condition (C). As in Tricomi's proof with the classical Rodrigues' type formula, these authors omit the assumption of orthogonality and determine all sequences $\{P_n(x)\}$, $P_n(x)$ of degree n, such that

$$P_n(x) = [w(x)]^{-1}\Delta^n[X(x)X(x - 1) \cdots X(x - n + 1)w(x - n)], \quad (3.1)$$

where $X(x)$ is a polynomial and both X and w are independent of n.

Assuming (3.1) produces polynomials, they determine a necessary form of the function w (which they express in terms of gamma functions). They then show that if the degree, k, of $X(x)$ exceeds 2, the degree of $P_2(x)$ will exceed 2. Thus $k \leq 2$ and by analysis of the three possible cases, they show that there are essentially only three classes of polynomials generated by the finite difference analogue of Rodrigues' formula.

For $k = 0$, the corresponding polynomials are the *Charlier polynomials* (VI-§1). For the monic polynomials, the resulting formula is

$$C_n^{(a)}(x) = (-1)^n a^{-x}\Gamma(x + 1)\Delta^n\left[\frac{a^x}{\Gamma(x - n + 1)}\right]. \qquad (3.2)$$

For $k = 1$, there are two forms of the function w possible. The first leads to Charlier polynomials again. The second produces the formula

$$m_n(x; \beta, c) = \frac{\Gamma(x + 1)}{\Gamma(x + \beta)} c^{-x-n} \Delta^n \left[\frac{c^x \Gamma(x + \beta)}{\Gamma(x - n + 1)} \right]. \tag{3.3}$$

These polynomials have the hypergeometric representation

$$m_n(x; \beta, c) = (\beta)_n {}_2F_1(-n, -x; \beta; 1 - c^{-1}). \tag{3.4}$$

The special case,

$$k_n(x) = p^n (n!)^{-1} m_n(x; -N, -p/q), \qquad n = 0, 1, \ldots, N, \tag{3.5}$$

was first studied by Krawtchouk [1]. Krawtchouk's polynomials are not positive-definite orthogonal polynomials but satisfy the finite orthogonality relation, for $p > 0$, $q > 0$, $p + q = 1$, N a positive integer,

$$\sum_{i=0}^{N} k_m(i) k_n(i) \binom{N}{i} p^i q^{N-i} = \binom{N}{n} p^n q^n \delta_{mn}, \qquad m, n = 0, 1, \ldots, N. \tag{3.6}$$

Thus the corresponding functional has a finite supporting set.

For additional properties of the Krawtchouk polynomials, see Eagleson [2], Erdélyi [1, vol. 2], Greenleaf [1], Karlin [2], [3], Karlin and McGregor [3], [5], Lesky [3], and Szegö [5]. Greenleaf and Karlin and McGregor discuss applications.

The general case of (3.4) was first studied by J. Meixner [1]. For $0 < c < 1$, $\beta > 0$, they form a positive-definite OPS being orthogonal with respect to a distribution step function whose jumps are

$$j(x; \beta, c) = \frac{c^x (\beta)_x}{x!} \qquad \text{at } x = 0, 1, 2, \ldots.$$

These polynomials are discussed in greater detail in VI-§3.

For $k = 2$, the corresponding Rodrigues' type formula can be put in the form

$$p_n(x; \alpha, \beta, \gamma) = \frac{\Gamma(x + 1)\Gamma(x + \gamma)}{n! \, \Gamma(x + \alpha)\Gamma(x + \beta)} \Delta^n \left[\frac{\Gamma(x + \alpha)\Gamma(x + \beta)}{\Gamma(x - n + 1)\Gamma(x - n + \gamma)} \right]. \tag{3.7}$$

These polynomials have the generalized hypergeometric form

$$p_n(x; \alpha, \beta, \gamma) = \frac{(\alpha)_n (\beta)_n}{n!} {}_3F_2(-n, -x, \alpha + \beta - \gamma + n; \alpha, \beta; 1). \tag{3.8}$$

First mentioned by Hahn in this generality, they include the special *discrete Tchebichef polynomials*,

$$t_n(x) = p_n(x; 1, 1 - N, 1 - N) \qquad (N \text{ a fixed positive integer}), \quad (3.9)$$

As with the Krawtchouk polynomials, these polynomials satisfy a finite orthogonality relation,

$$\sum_{k=0}^{N-1} t_m(k)t_n(k) = (2n + 1)^{-1}N(N^2 - 1^2)(N^2 - 2^2)\cdots(N^2 - n^2)\delta_{mn},$$

$$m, n = 0, 1, \ldots, N - 1. \quad (3.10)$$

They were introduced by Tchebichef [2] for use in least squares data fitting (for example, see Hildebrand [1] where they are referred to as *Gram polynomials*). For additional properties and references, see Szegö [5], Erdélyi [1, vol. 2] and Morrison [3] where applications are given.

Another special case of $p_n(x; \alpha, \beta, \gamma)$ is *Pasternak's polynomial* which is obtained by choosing $\alpha = 1$, $\beta = \gamma = \lambda + 1$ (Pasternak [1]). In turn, Pasternak's polynomial is a generalization of a polynomial studied by H. Bateman [1], the latter corresponding to $\lambda = 0$. These polynomials are discussed briefly in VI-§8.

In the general case, the polynomials are orthogonal with respect to the distribution step function whose jumps are

$$j(x; \alpha, \beta, \gamma) = \frac{(\alpha)_x(\beta)_x}{(\gamma)_x x!} \qquad \text{at } x = 0, 1, 2, \ldots$$

In order that the distribution function have finite moments, however, it is necessary that α or β be a negative integer since otherwise the infinite series corresponding to the pth moment diverges for p sufficiently large. Thus the distribution function has a finite spectrum.

For the general case, it may be preferable to use the notation of Karlin and McGregor [5] who have given the most extensive single treatment of these, the *Hahn polynomials*. These authors write

$$Q_n(x) = Q_n(x; \alpha, \beta, N)$$
$$= \frac{(\alpha + 1)_n(1 - N)_n}{n!}p_n(x; \alpha + 1, 1 - N, 1 - \beta - N), \quad (3.11)$$

where, for orthogonality purposes, $\alpha > -1$, $\beta > -1$, and N is a positive integer.

They write the orthogonality property as

$$\sum_{x=0}^{N-1} Q_m(x)Q_n(x)\rho(x) = \frac{\delta_{mn}}{\pi_n}, \qquad m, n = 0, 1, \ldots, N-1, \quad (3.12)$$

and obtain in addition the "dual orthogonality" relation

$$\sum_{n=0}^{N-1} Q_n(x)Q_n(y)\pi_n = \frac{\delta_{xy}}{\rho(x)}, \qquad x, y = 0, 1, \ldots, N-1. \quad (3.13)$$

Here

$$\pi_n = \pi_n(\alpha, \beta, N) = \frac{\binom{N-1}{n}}{\binom{N+\alpha+\beta+n}{n}} \cdot \frac{\Gamma(\beta+1)}{\Gamma(\alpha+1)\Gamma(\alpha+\beta+1)}$$

$$\cdot \frac{\Gamma(n+\alpha+1)\Gamma(n+\alpha+\beta+1)}{\Gamma(n+\beta+1)\Gamma(n+1)} \cdot \frac{(2n+\alpha+\beta+1)}{(\alpha+\beta+1)}$$

$$\rho(x) = \rho(x; \alpha, \beta, N) = \frac{\binom{\alpha+x}{x}\binom{\beta+N-1-x}{N-1-x}}{\binom{N+\alpha+\beta}{N-1}}.$$

For additional properties, applications, and further references, see Erdélyi [1, vol. 2], Karlin and McGregor [5], [7], [8], Lee [1], Levit [1], and Wilson [1].

4 The Meixner Class of Orthogonal Polynomials

Consider two formal power series

$$A(w) = \sum_{n=0}^{\infty} a_n w^n, \qquad a_0 \neq 0,$$

$$G(w) = \sum_{n=1}^{\infty} g_n w^n, \qquad g_1 \neq 0.$$

It is easy to verify that the coefficients $f_n(x)$ in the formal power series

$$A(w)e^{xG(w)} = \sum_{n=0}^{\infty} f_n(x)w^n \quad (4.1)$$

are polynomials in x of degree n with leading coefficients $(n!)^{-1}a_0 g_1^n$.

Polynomial sequences $\{f_n(x)\}$ having generating functions of the form

(4.1) are called, variously, sets of "A-type zero" (Sheffer [1], [2]), "generalized Appell type" (Erdélyi [1]), and "Sheffer type" (Boas and Buck [1]). The first investigation of this class of polynomial generating functions seems to be that of J. Meixner [1].

Meixner considered the problem of determining all positive-definite OPS having generating functions of the form (4.1). Choosing, without loss of generality, $a_0 = g_1 = 1$, and considering the corresponding monic polynomials

$$P_n(x) = n! f_n(x),$$

Meixner then proved that $\{P_n(x)\}$ has a generating function of the form (4.1) if and only if it satisfies a recurrence formula of the form

$$P_{n+1}(x) = [x - (dn + f)]P_n(x) - n(gn + h)P_{n-1}(x), \qquad n \geq 0, \quad (4.2)$$

where $g \geq 0$, $g + h > 0$, d and f are real.

(f can be considered arbitrary since altering its value amounts only to a translation in x.)

Meixner then showed that there are five distinct classes of positive-definite OPS determined by (4.2).

(A) $d = 0$, $g = 0$. With the choice $f = 0$, the recurrence becomes

$$P_{n+1}(x) = xP_n(x) - hnP_{n-1}(x) \qquad (h > 0),$$

so that $P_n(x) = (h/2)^{n/2} H_n(cx)$, $c = (2h)^{-1/2}$, where $H_n(x)$ is the Hermite polynomial. The corresponding generating function is of course the classical one, (2.39).

(B) $d \neq 0$, $d^2 - 4g = 0$. Choose $f = (h + g)g^{-1/2}$ so the recurrence formula becomes

$$P_{n+1}(x) = [x - (2n + \frac{h}{g} + 1)g^{1/2}]P_n(x) - gn(n + \frac{h}{g})P_{n-1}(x).$$

It follows that in this case,

$$P_n(x) = (-1)^n g^{n/2} n! L_n^{(h/g)}(xg^{-1/2}),$$

where $L_n^{(\alpha)}(x)$ denotes the Laguerre polynomial. Once again, the generating function is the classical one, (2.38).

(C) $d \neq 0$, $g = 0$. This time choose $f = h/d$ to obtain the recurrence formula,

$$P_{n+1}(x) = [x - d(n + hd^{-2})]P_n(x) - hnP_{n-1}(x),$$

and conclude that $P_n(x) = d^n C_n^{(a)}(x/d)$, $a = hd^{-2}$, where $C_n^{(a)}(x)$ is the Charlier polynomial. The generating function in this case corresponds to the usual one which was given in I-(1.11).

(D) $d^2 - 4g > 0$. Here choose

$$f = \frac{2(g + h)}{d + \rho} \quad \text{where} \quad \rho = (d^2 - 4g)^{1/2}.$$

In this case, a new class of orthogonal polynomials is obtained. These are the polynomials (3.4) later obtained by Hahn. For the orthogonality relations, Meixner obtains the distribution step function whose jumps are

$$j(x) = \left(\frac{d - \rho}{d + \rho}\right)^n \frac{(1 + h/g)_n}{n!} \quad \text{at } x = \rho n, \quad n = 0, 1, 2, \ldots . \quad (4.3)$$

The corresponding generating function is

$$[1 + (d + \rho)w/2]^{x/\rho}[1 + (d - \rho)w/2]^{-x/\rho - h/g - 1} = \sum_{n=0}^{\infty} P_n(x)\frac{w^n}{n!}. \quad (4.4)$$

These polynomials will be discussed in more detail in VI-§3.

(E) $d^2 - 4g < 0$. Once again, a new class of orthogonal polynomials is obtained. Choosing $f = 0$ and writing $\sigma = (4g - d^2)^{1/2}$, these polynomials are orthogonal over $(-\infty, \infty)$ with respect to the weight function

$$w(x) = \exp\{-\frac{x}{\sigma} \arg(z_0)\}\Gamma\left(z_1 + \frac{ix}{\sigma}\right)\Gamma\left(\bar{z}_1 - \frac{ix}{\sigma}\right), \quad (4.5)$$

where

$$z_0 = -\frac{d + i\sigma}{d - i\sigma}, \quad z_1 = \frac{g + h}{2\sigma g}(\sigma + id),$$

$$-\pi < \arg(z_0) < \pi.$$

Meixner's generating function is

$$\left(\frac{1 + \alpha t}{1 + \beta t}\right)^{-ix/\sigma}\left[\frac{(1 + \beta t)^\alpha}{(1 + \alpha t)^\beta}\right]^{i(g+h)/g\sigma} = \sum_{n=0}^{\infty} P_n(x)\frac{t^n}{n!}, \quad (4.6)$$

where $2\alpha = d + i\sigma$, $\beta = \bar{\alpha}$.

These polynomials will also be discussed in more detail in VI-§3.

In the special cases (A) and (E) with $d = 0$, $\{P_n(x)\}$ is a symmetric OPS. It thus follows that $\{P_{2n}(x^{1/2})\}$ is an OPS and $\{x^{-1/2}P_{2n+1}(x^{1/2})\}$ is the corresponding KPS with K-parameter 0 (see I-§8).

It was observed by W. A. Al-Salam [5] that it can happen that $\{P_n(x)\}$ is not an OPS but nevertheless $\{P_{2n}(x^{1/2})\}$ and $\{x^{-1/2}P_{2n+1}(x^{1/2})\}$ are both OPS. Al-Salam thus considered the problem of determining all polynomial sequences generated by (4.1) such that

   (i) $f_n(-x) = (-1)^n f_n(x)$,
   (ii) either $\{F_n(x)\} = \{f_{2n}(x^{1/2})\}$ or $\{G_n(x)\} = \{x^{-1/2}f_{2n+1}(x^{1/2})\}$ is an OPS.

Al-Salam first proved that if either $\{F_n(x)\}$ or $\{G_n(x)\}$ is an OPS, then so is the other. He then shows that in addition to the two cases noted above, there are essentially only three other cases in which $\{F_n(x)\}$ and $\{G_n(x)\}$ are OPS.

(F) $A(w) = 1$, $G(w) = \lambda(w)$ where λ denotes that inverse of the Jacobian elliptic function, $sn(x)$, that vanishes at the origin. For the corresponding generating function, Al-Salam obtains

$$x^{-1/2}f_{2n+1}(x^{1/2}) = A_n(x), \qquad x^{-1}f_{2n}(x^{1/2}) = B_{n-1}(x),$$

where $\{A_n(x)\}$ and $\{B_n(x)\}$ are certain OPS studied by Stieltjes [2] and Carlitz [6] (see VI-§9).

Strictly speaking, the above does not conform exactly to the condition (ii) of the original problem. However, Al-Salam shows that there is the alternate generating function with

$$A(w) = \lambda'(w), \qquad G(w) = \lambda(w),$$

for which $F_n(x) = A_n(x)$ and $G_n(x) = B_n(x)$.
(G) $A(w) = (1 - w^2)^{-1/2}$, $G(w) = \lambda(w)$. For this case,

$$F_n(x) = C_{2n}(x^{1/2}), \qquad G_n(x) = x^{-1/2}D_{2n+1}(x^{1/2}),$$

where $\{C_n(x)\}$ and $\{D_n(x)\}$ are certain symmetric OPS also studied by Stieltjes and by Carlitz.
(H) $A(w) = (1 - k^2 w^2)^{-1/2}$, $G(w) = \lambda(w)$. This is a companion to (G) for we have

$$F_n(x) = D_{2n}(x^{1/2}), \qquad G_n(x) = x^{-1/2}C_{2n+1}(x^{1/2}).$$

The polynomials $C_n(x)$ and $D_n(x)$ will also be discussed in VI-§9.

5   Other Classes of Orthogonal Polynomials

Polynomial sequences $\{p_n(x)\}$ having generating functions of the form

$$A(w)B(xw) = \sum_{n=0}^{\infty} p_n(x)w^n \qquad (5.1)$$

were first studied by W. Brenke [1]. These polynomials are simple in structure since, if

$$A(w) = \sum_{n=0}^{\infty} a_n w^n, \qquad B(w) = \sum_{n=0}^{\infty} b_n w^n,$$

then

$$p_n(x) = \sum_{k=0}^{n} a_{n-k} b_k x^k, \qquad n = 0, 1, 2, \ldots \qquad (5.2)$$

In particular, the leading coefficient of $p_n(x)$ is $a_0 b_n$ so $p_n(x)$ is of degree n provided $a_0 \neq 0$, $b_n \neq 0$.

All OPS having generating functions of the form (5.1) were determined by Chihara [6]. They are, essentially, the following:

(A) the Laguerre polynomials;

(B) the generalized Hermite polynomials, $H_n^{(\mu)}(x)$;

(C) the Wall polynomials, $W_n(x; b, q)$ (see VI-§11);

(D) a symmetric set of polynomials, $Y_n(x; b, q)$, defined by

$$Y_{2n}(x) = W_n(x^2; b, q), \qquad Y_{2n+1}(x) = x W_n(x^2; bq, q)$$

(see VI-§11);

(E) the generalized Stieltjes-Wigert polynomials, $S_n(x; p, q)$ (see VI-§2);

(F) a set of symmetric polynomials, $T_n(x; p, q)$, defined by

$$T_{2n}(x) = S_n(x^2; p, q), \qquad T_{2n+1}(x) = q^{-n} x S_n(qx^2; pq, q)$$

(see VI-§2);

(G) the q-polynomials of Al-Salam and Carlitz, $U_n^{(a)}(x)$ and $V_n^{(a)}(x)$ (see VI-§10);

(H) a set of (non-symmetric) polynomials, $Z_n(x; b, q)$, related to the Wall polynomials by

$$Z_{2n}(x) = W_n(x^2; b, q) + (1 - q^n)x W_{n-1}(x^2; bq, q)$$

$$Z_{2n+1}(x) = x W_n(x^2; bq, q) + (1 - b)W_n(x^2; b, q).$$

(see VI-§11)

In an earlier study, J. Geronimus [4] studied polynomials of the form

$$G_n(x) = \sum_{k=0}^{n} a_{n-k} b_k \pi_k(x) \tag{5.3}$$

where $\pi_0(x) = 1$, $\pi_k(x) = (x - x_1) \cdots (x - x_k)$.
Geronimus obtains necessary and sufficient conditions on the sequences $\{a_n\}$, $\{b_n\}$, and $\{x_n\}$ in order that $\{G_n(x)\}$ form an OPS. Although he does not determine all such OPS explicitly, Geronimus does note a large number of previously studied cases that can be put in the form (5.3). In particular, the Charlier, Meixner, and Stieltjes-Wigert polynomials belong to this class (as do, obviously, all the previously mentioned cases which have Brenke type generating functions).

An interesting subclass of these orthogonal polynomials is also discussed by Geronimus. Let $\sum_{k=0}^{\infty} d_k z^k$ be the formal reciprocal power series of $A(z) = \sum a_k z^k$ with $a_0 d_0 = 1$. Let

$$J_i = \sum_{k=i}^{\infty} \frac{d_{k-1}}{b_{k-1} \pi'_k(x_i)}, \qquad i = 1, 2, 3, \ldots \tag{5.4}$$

Geronimus then proves: Let the x_i be distinct real numbers and let $\{G_n(x)\}$ be a positive-definite OPS. Finally, suppose the series (5.4) all converge absolutely. Then the $G_n(x)$ are orthogonal with respect to the distribution step function whose spectrum is $\{x_1, x_2, x_3, \ldots\}$ and whose jump at x_i is J_i.

Geronimus illustrates this theorem with examples which include the polynomials of Charlier, Meixner, and Stieltjes-Wigert. The determination of all OPS belonging to even this subclass remains an intriguing problem.

Analogous to Bochner's study of the second order Sturm-Liouville differential equation, analyses have been made of the difference equation

$$a(x)\Delta^2 y(x) + b(x)\Delta y(x) + [c(x) + \lambda_n]y(x + 1) = 0 \tag{5.5}$$

by Lancaster [1] and Lesky [4], [5].
Their work shows that if $a(x)$, $b(x)$ and $c(x)$ are polynomials independent of n, then the only orthogonal polynomial solutions are the Charlier and Meixner polynomials and two finite sets, the Krawtchouk and the Hahn polynomials.

Krall and Sheffer [1] have classified orthogonal polynomials in terms of differential equations of the form

$$F[y] = \sum_{k=1}^{\infty} M_k(x)y^{(k)}(x) = \lambda y(x) \tag{5.6}$$

where $M_k(x)$ is a polynomial of degree k. They prove that for each OPS $\{P_n(x)\}$ there is a corresponding operator (5.6) such that $F[P_n(x)] = \lambda_n P_n(x)$, the eigenvalues λ_n are expressed simply in terms of the coefficients in the $M_k(x)$ and $\lambda_j \neq \lambda_k$ for $j \neq k$.

They then classify OPS according to the maximum degree that appears among the coefficient polynomials $M_k(x)$. Their results concerning this classification are somewhat involved and do not lend themselves to convenient summarization here.

A sequence of polynomials $\pi_n(x)$ is called an *Appell sequence* if

$$\pi_n'(x) = \pi_{n-1}(x), \qquad n = 1, 2, 3, \ldots; \pi_{-1}(x) = 0.$$

It follows from (2.21) that $\{(n!)^{-1} H_n(x)\}$ is an Appell sequence and it follows from Hahn's characterization of the classical orthogonal polynomials as the only ones whose derivatives are also orthogonal that the Hermite polynomials are the only ones that are both Appell and orthogonal. This is one of the oldest characterizations of a class of orthogonal polynomials and one of the most frequently rediscovered as well. Credit for the first discovery seems to be due to A. Angelescu [1].

Another class of polynomial sequences which turns out to contain only one OPS consists of sequences $\{P_n(x)\}$ where

$$P_n(x) = {}_{p+1}F_q(-n, a_1, \ldots, a_p; b_1, \ldots, b_q; x),$$

$p \geq 0$, $q \geq 1$, and the a_i and b_j are independent of n and x. It was shown by N. Abdul-Halim and W. A. Al-Salam [1] that the only OPS of this type is essentially the Laguerre polynomial sequence for which there is the well known representation

$$L_n^{(\alpha)}(x) = \binom{n + \alpha}{n} {}_1F_1(-n; \alpha + 1; x).$$

Many orthogonal polynomials can be represented by generalized hypergeometric functions if the various parameters are allowed to depend upon n and/or x and a number of characterizations of such OPS in terms of the particular type of hypergeometric representation are given by N. A. Al-Salam [1].

W. A. Allaway [1] has determined all polynomial sequences $\{A_n(x)\}$ which are orthogonal with respect to a weight function w such that $A_n(\cos\theta)w(\cos\theta)$ has a Fourier sine expansion of a certain form. This generalizes a property of ultraspherical polynomials due to Szegö [4].

Meixner [2] has shown that the only OPS $\{S_n(x)\}$ with the property $S_n(m) = S_m(n)$ for all nonnegative integers m and n are the Charlier and the two Meixner polynomial sets of (4.4) and (4.6).

CHAPTER VI

Some Specific Systems of Orthogonal Polynomials

1 The Charlier Polynomials

We will define the monic *Charlier polynomials*, $C_n^{(a)}(x)$, by the generating function

$$e^{-aw}(1 + w)^x = \sum_{n=0}^{\infty} C_n^{(a)}(x)\frac{w^n}{n!}, \qquad a \neq 0. \tag{1.1}$$

(See I-§1 where $(n!)^{-1} C_n^{(a)}(x)$ is denoted by $P_n(x)$.)
    The explicit representation is

$$C_n^{(a)}(x) = \sum_{k=0}^{n} \binom{n}{k}\binom{x}{k}k!\,(-a)^{n-k}, \tag{1.2}$$

and the orthogonality relation is

$$\int_0^\infty C_m^{(a)}(x)C_n^{(a)}(x)\,d\psi^{(a)}(x) = a^n n!\,\delta_{mn}, \tag{1.3}$$

where $\psi^{(a)}$ is the step function whose jumps are

$$d\psi^{(a)}(x) = \frac{e^{-a}a^x}{x!} \qquad \text{at } x = 0, 1, 2, \dots$$

Thus the positive-definite case occurs for $a > 0$ and in this case, $d\psi^{(a)}(x)$ is the Poisson distribution of probability theory.
    The recurrence formula is

$$C_{n+1}^{(a)}(x) = (x - n - a)C_n^{(a)}(x) - an C_{n-1}^{(a)}(x). \tag{1.4}$$

The Charlier polynomials can be expressed in terms of Laguerre polynomials (cf. (1.2) and V-(2.11)):

$$C_n^{(a)}(x) = n! \, L_n^{(x-n)}(a). \tag{1.5}$$

Reflecting the relationship of the binomial distribution to the Poisson distribution, the Charlier polynomials are limiting cases of the Krawtchouk polynomials, V-(3.5). Specifically, put $pN = a$, $q = 1 - p$. Then for fixed a, x and n,

$$\lim_{N \to \infty} k_n(x) = (n!)^{-1} C_n^{(a)}(x) \tag{1.6}$$

(Szegö [5], p. 36; cf. also Young [1]).

The generating function (1.1) is of course of Meixner type. The Charlier polynomials also belong to the Hahn class of orthogonal polynomials and the finite difference Rodrigues' type formula was given in V-(3.2). There is also the simple difference relation

$$\Delta C_n^{(a)}(x) = n C_{n-1}^{(a)}(x). \tag{1.7}$$

In connection with (1.4), this yields the second order difference equation

$$a\Delta^2 C_n^{(a)}(x) - (x + 1 - a - n)\Delta C_n^{(a)}(x) + n C_n^{(a)}(x) = 0. \tag{1.8}$$

The Charlier polynomials were introduced by Charlier [1]. Szegö [5] refers to them as the Poisson-Charlier polynomials and Erdélyi [1] uses the notation

$$c_n(x; a) = (-1)^n a^{-n} C_n^{(a)}(x) \tag{1.9}$$

This non-monic form of the Charlier polynomials may be preferable for some purposes. For one thing, there is then the simple symmetry relation

$$c_n(x; a) = c_x(n; a). \tag{1.10}$$

From this there follows the "dual orthogonality" relation

$$\sum_{x=0}^{\infty} c_x(m; a) c_x(n; a) \frac{e^{-a} a^x}{x!} = a^{-n} n! \, \delta_{mn}. \tag{1.11}$$

The general concept of dual orthogonality relations for discrete orthogonal systems is discussed by Atkinson [1] and Eagleson [1]. The latter characterizes discrete orthogonal systems for which dual orthogonal systems exist in terms of completeness.

For additional properties of the Charlier polynomials, see Erdélyi [1], Geronimus [1], and Karlin [2], [3] and Karlin and McGregor [4] for an application.

It is of some interest to note that $\{C_n^{(a)}(x-1)\}$ is the monic KPS with K-parameter 0 corresponding to $\{C_n^{(a)}(x)\}$. Defining $\{D_n^{(a)}(x)\}$ by

$$D_{2m}^{(a)}(x) = C_m^{(a)}(x^2), \qquad D_{2m+1}^{(a)}(x) = x C_m^{(a)}(x^2 - 1), \qquad (1.12)$$

we obtain the symmetric OPS with respect to the distribution function, $\phi^{(a)}$, whose jumps are

$$d\phi^{(a)}(x) = d\psi^{(a)}(x^2) = \frac{e^{-a} a^{x^2}}{(x^2)!}, \qquad x = \pm 0, \pm 1, \pm\sqrt{2}, \pm\sqrt{3}, \ldots,$$

where we have written ± 0 to indicate the jump at 0 is to be counted twice.

The recurrence formula has the simple form

$$D_n^{(a)}(x) = x D_{n-1}^{(a)}(x) - \gamma_n D_{n-2}^{(a)}(x), \qquad n \geq 1,$$
$$\gamma_{2m} = a, \qquad \gamma_{2m+1} = m. \qquad (1.13)$$

The Hamburger moment problems associated with the Charlier polynomials and with the associated symmetric polynomials are determined. This is quickly verified by Carleman's criterion that the Hamburger moment problem associated with the positive-definite OPS $\{P_n(x)\}$ whose recurrence formula is IV-(1.1) is determined if

$$\sum_{n=2}^{\infty} \lambda_n^{-1/2} = \infty \qquad (1.14)$$

(see Shohat and Tamarkin [1, p. 59]).

2 The Stieltjes-Wigert Polynomials

Stieltjes [2, pp. 507-508] introduced the weight function

$$w(x) = k\pi^{-1/2} x^{-k^2 \log x} = k\pi^{-1/2} \exp\{-k^2 \log^2 x\}, \qquad 0 < x < \infty, \quad (2.1)$$

(with $k = 1$) as an example leading to an indeterminate Stieltjes moment problem (cf. II-§6). He also provided the continued fraction expansion,

$$\frac{1}{\sqrt{\pi}} \int_0^\infty \frac{x^{-\log x}}{z + x} dx = \frac{1|}{|a_1 z} + \frac{1|}{|a_2} + \frac{1|}{|a_3 z} + \frac{1|}{|a_4} + \cdots, \qquad (2.2)$$

where

$$
\begin{aligned}
a_{2m} &= q^m(1-q)\cdots(1-q^{m-1}) \\
a_{2m+1} &= q^{m+1/2}[(1-q)\cdots(1-q^m)]^{-1}.
\end{aligned}
\qquad q = e^{-1/2},
$$

The continued fraction in (2.2) is a Stieltjes type fraction (S-fraction). The "even part" of this continued fraction is a J-fraction whose sequence of convergents is the same as the convergents of even order of the S-fraction, and its partial denominators are the orthogonal polynomials with respect to the weight function (2.1). By the method of "contraction" (see for example, Wall [2]), the even part of (2.2) can be obtained and the three term recurrence formula for these orthogonal polynomials can be determined.

S. Wigert [1] obtained for general $k > 0$ the explicit representation for the corresponding ortho*normal* polynomials as

$$
p_n(x) = (-1)^n q^{(2n+1)/4}[q]_n^{-1/2} \sum_{j=0}^{\infty} \begin{bmatrix} n \\ j \end{bmatrix} q^{j^2}(-q^{1/2}x)^j, \qquad (2.3)
$$

where now, $q = \exp[-(2k^2)^{-1}]$, and the symbol $[a]_n$ is defined by

$$
[a]_0 = 1, \qquad [a]_n = (1-a)(1-aq)\cdots(1-aq^{n-1}) \qquad \text{for } n \geqq 1. \quad (2.4)
$$

Also, we have used Gauss' symbol,

$$
\begin{bmatrix} n \\ j \end{bmatrix} = \frac{[q]_n}{[q]_{n-j}[q]_j}. \qquad (2.5)
$$

More generally, we can introduce a second parameter p, $0 \leqq p < 1$, and define for $0 < x < \infty$,

$$
\begin{aligned}
\rho(x) \equiv \rho(x; p, q) &= [p]_\infty \prod_{m=1}^{\infty} (1 - pq^{m-3/2}x^{-1})w(x) \\
&= [p]_\infty \sum_{n=0}^{\infty} \frac{q^{n(n-2)/2}(px^{-1})^n}{[q]_n} w(x),
\end{aligned}
\qquad (2.6)
$$

where we have written

$$
[p]_\infty = \prod_{m=0}^{\infty} (1 - pq^m).
$$

Then the *generalized Stieltjes-Wigert polynomials,*

$$S_n(x; p, q) = (-1)^n q^{-n(2n+1)/2} [p]_n \sum_{m=0}^{n} \begin{bmatrix} n \\ m \end{bmatrix} \frac{q^{m^2}(-q^{1/2}x)^m}{[p]m}, \quad (2.7)$$

are the monic polynomials that are orthogonal with respect to the weight function *rho*(x) on (0, *inf*) (see Chihara [6]).

We have

$$S_n(x; 0, q) = q^{-(2n+1)^2/4} [q]_n^{1/2} p_n(x). \quad (2.8)$$

The recurrence formula for $\{S_n(x)\} = \{S_n(x; p, q)\}$ is

$$
\begin{aligned}
S_{n+1}(x) = \{x + [p + q - (1 + q)q^{-n}]q^{-n-3/2}\} S_n(x) \\
- (1 - q^n)(1 - pq^{n-1})q^{-4n} S_{n-1}(x).
\end{aligned}
\quad (2.9)
$$

The explicit orthogonality relation is

$$\int_0^\infty S_m(x) S_n(x) \rho(x)\, dx = [p]_n [q]_n q^{-(2n+1)^2/2} \delta_{mn}. \quad (2.10)$$

The moments are

$$\mu_n = [p]_n q^{-(n+1)^2/2}. \quad (2.11)$$

Thus $\{\mu_n\}$ satisfies Hahn's condition E (see V-§3) and hence the generalized Stieltjes-Wigert polynomials belong to the Hahn class of orthogonal polynomials. They also belong to the class of orthogonal polynomials having Brenke type generating functions:

$$A(w) B(xw) = \sum_{n=0}^{\infty} \frac{q^{n^2}}{[p]_n [q]_n} S_n(x; p, q) w^n, \quad (2.12)$$

where

$$A(w) = \sum_{n=0}^{\infty} \frac{(-q^{-1/2})^n w^n}{[q]_n}, \qquad B(w) = \sum_{n=0}^{\infty} \frac{q^{n^2} w^n}{[p]_n [q]_n}.$$

The observation that $\{q^{-n} S_n(qx; pq, q)\}$ is the monic KPS with K-parameter 0 corresponding to $\{S_n(x; p, q)\}$ leads to setting

$$T_{2n}(x; p, q) = S_n(x^2; p, q), \qquad T_{2n+1}(x) = q^{-n} x S_n(qx^2; pq, q). \quad (2.13)$$

Then $\{T_n(x)\} = \{T_n(x;p,q)\}$ is the monic symmetric OPS with respect to the weight function $|x|\rho(x^2)$ on $(-\infty, \infty)$. The recurrence formula is

$$T_n(x) = xT_{n-1}(x) - \gamma_n T_{n-2}(x), \qquad n \geqq 1,$$
$$\gamma_{2n} = (1 - pq^{n-1})q^{-2n+1/2}, \qquad \gamma_{2n+1} = (1 - q^n)q^{-2n-1/2}. \tag{2.14}$$

They have the Brenke type generating function

$$A(w^2)[B(x^2w^2;p,q) + \frac{q^{1/4}xw}{1-p}B(qx^2w^2;pq,q)] = \sum_{n=0}^{\infty} \frac{q^{n^2/4}}{C_n}T_n(x)w^n, \tag{2.15}$$

where $C_{2k} = [p]_k[q]_k$ and $C_{2k+1} = [p]_{k+1}[[q]_k$.

It has been noted that the Stieltjes moment problem corresponding to the ordinary Stieltjes-Wigert polynomials is indeterminate (hence so is the Hamburger moment problem indeterminate). Stieltjes' example also serves to show that the Stieltjes moment problem corresponding to the generalized Stieltjes-Wigert polynomials is indeterminate. It is then readily concluded that the Hamburger moment problem associated with the above symmetric OPS is also indeterminate.

For other properties of the Stieltjes-Wigert polynomials, see Chihara [10], Geronimus [4] and Szegö [3] where closely related polynomials which are orthogonal over the unit circle in the complex plane are also studied. The latter are also discussed by Carlitz [1], [4].

3   The Meixner Polynomials

The two general classes of orthogonal polynomials corresponding to cases D and E of V-§4 can be characterized by the recurrence formula

$$P_{n+1}(x) = [x - (dn + f)]P_n(x) - n(gn + h)P_{n-1}(x), \tag{3.1}$$

where

$$g > 0, \qquad g + h > 0, \qquad d \text{ and } f \text{ real}, \qquad d^2 - 4g \neq 0.$$

The resulting positive-definite orthogonal polynomials are of distinctly different character according as $d^2 - 4g$ is positive or negative.

(A) *Meixner polynomials of the first kind.* Consider first the case, $d^2 - 4g > 0$, and write

$$\rho = \sqrt{d^2 - 4g}.$$

We can assume without loss of generality that

$$d > 0 \quad \text{and} \quad f = \frac{2(g + h)}{d + \rho}.$$

We then write

$$c = \frac{d - \rho}{d + \rho}, \qquad \beta = 1 + \frac{h}{g},$$

and set

$$m_n(x; \beta, c) = m_n(x) = \left(\frac{c - 1}{c\rho} \right)^n P_n(\rho x).$$

The recurrence formula (3.1) then becomes

$$cm_{n+1}(x) = [(c - 1)x + (1 + c)n + c\beta]m_n(x) - n(n + \beta - 1)m_{n-1}(x),$$

$$n \geq 0, \quad (3.2)$$

The original conditions for positive-definite orthogonality require that $0 < c < 1$ and $\beta > 0$. However, we will now consider the *Meixner polynomials of the first kind* to be the polynomials defined by (3.2) with the restrictions

$$c \neq 0, 1 \quad \text{and} \quad \beta \neq 0, -1, -2, \ldots$$

Meixner's generating function can now be written

$$\left(1 - \frac{w}{c}\right)^x (1 - w)^{-x - \beta} = \sum_{n=0}^{\infty} m_n(x; \beta, c) \frac{w^n}{n!}. \quad (3.3)$$

The corresponding orthogonality relation is, for $0 < |c| < 1$,

$$\sum_{k=0}^{\infty} m_n(k; \beta, c) m_p(k; \beta, c) \frac{c^k (\beta)_k}{k!} = (1 - c)^{-\beta} c^{-n} n! (\beta)_n \delta_{np}. \quad (3.4)$$

This can be verified from the generating function in exactly the same manner as with the Charlier polynomials (cf. I-§1). The generating function also yields the explicit representation

$$m_n(x; \beta, c) = (-1)^n n! \sum_{k=0}^{n} \binom{x}{k} \binom{-x - \beta}{n - k} c^{-k}. \quad (3.5)$$

The easily verified identity

$$m_n(-x - \beta; \beta, a^{-1}) = a^n m_n(x; \beta, a) \qquad (3.6)$$

can be used to derive from (3.4) the orthogonality relation for $|c| > 1$:

$$\sum_{k=0}^{\infty} m_n(x_k; \beta, c) m_p(x_k; \beta, c) \frac{c^{-k}(\beta)_k}{k!} = \left(\frac{c}{c-1}\right)^\beta c^{-n} n! (\beta)_n \delta_{pn}, \qquad (3.7)$$

where

$$x_k = -k - \beta, \qquad k = 0, 1, 2, \ldots$$

Corresponding to the excluded case, $c = 1$, there is the readily verified limit

$$\lim_{c \to 1} m_n\left(\frac{x}{1-c}; \beta, c\right) = n! L_n^{(\beta-1)}(x). \qquad (3.8)$$

The Meixner polynomials were later studied independently by Feldheim [1] who obtained a number of additional properties. In particular, he obtained the difference equations

$$\Delta m_n(x; \beta, c) = \left(1 - \frac{1}{c}\right) n m_{n-1}(x; \beta + 1, c),$$

$$c(x + \beta + 1)\Delta^2 m_n(x) + [(c - 1)(x - n + 1) + \beta c]\Delta m_n(x) \qquad (3.9)$$
$$-n(c - 1)m_n(x) = 0$$

and the finite difference Rodriques' type formula, V-(3.3), which correspond to the characterizations (A), (B) and (C) of Hahn's class of orthogonal polynomials (V-§3). Other relations include the hypergeometric representation V-(3.4) and generating functions including a bilinear one of "Hardy-Hille type." Asymptotic properties are also studied.

Additional properties of the Meixner polynomials can be found in Karlin [2], [3], Karlin and Szegö [1], Lesky [1] and Prizva [1]. The special case $\beta = 1$ had already been considered briefly by Stieltjes [2] and later in some detail by Gottlieb [1]. More recently, Carlitz [5] made an extensive study of this case from a different viewpoint. Specifically, Carlitz considers

$$A_n^{(\lambda)}(x) = (n!)^{-1} m_n(x; 1, \lambda^{-1}) \qquad (3.10)$$

and obtains the general orthogonality relation

$$\mathcal{L}[x^m A_n^{(\lambda)}(x)] = \frac{n! \, \lambda^n}{(1-\lambda)^n} \delta_{mn}, \qquad (3.11)$$

where the moments of $\mathcal{L}$ are the "Eulerian numbers," $R_k(\lambda)$, defined by

$$\frac{1-\lambda}{e^x - \lambda} = \sum_{k=0}^{\infty} R_k(\lambda) \frac{x^k}{k!}.$$

For the case, $|\lambda| > 1$, Carlitz also obtains the orthogonality relation (3.4) as well as the complex form

$$\int_{\alpha - i\infty}^{\alpha + i\infty} A_m^{(\lambda)}(x) A_n^{(\lambda)}(x) \frac{(-\lambda)^{-z} \, dz}{\sin \pi z} = \frac{2i \lambda^{n+1}}{1-\lambda} \delta_{mn}, \qquad -1 < \alpha < 0. \quad (3.12)$$

The Meixner polynomials are of importance for the description of certain Markov processes (birth and death processes) in a theory developed by Karlin and McGregor [3]. The special case $\beta = 1$ has also been applied in the error analysis of certain extrapolation problems by Duffin and Schmidt [1], N. Morrison [1], [3]. These latter writers refer to these polynomials as "discrete Laguerre polynomials" (note the limit (3.8)).

Duffin and Schmidt actually work with the kernel polynomials with K-parameter 0. In this connection, it is interesting to note that more generally, the kernel polynomials with K-parameter 0 corresponding to $\{m_n(x; \beta, c)\}$ are $m_n(x-1; \beta+1, c)$. Thus introducing the monic polynomials,

$$\hat{m}_n(x; \beta, c) = \left(\frac{c}{c-1}\right)^n m_n(x; \beta, c), \qquad (3.13)$$

we set

$$\begin{aligned} R_{2n}(x; \beta, c) &= \hat{m}_n(x^2; \beta, c) \\ R_{2n+1}(x; \beta, c) &= x \hat{m}_n(x^2 - 1; \beta + 1, c). \end{aligned} \qquad (3.14)$$

Then $\{R_n(x; \beta, c)\}$ is the monic OPS with respect to the distribution step function whose jumps are

$$\frac{c^{k^2} (\beta)_{k^2}}{(k^2)!} \qquad \text{at } k = \pm 0, \pm 1, \pm\sqrt{2}, \pm\sqrt{3}, \dots$$

where we write ± 0 to indicate that the jump at 0 is to be counted twice.

From the recurrence formula

$$\hat{m}_{n+1}(x) = \left[x - \frac{(c+1)n + \beta c}{1-c} \right] \hat{m}_n(x) - \frac{c}{(1-c)^2} n(n + \beta - 1) \hat{m}_{n-1}(x),$$

$$(3.15)$$

we obtain the recurrence

$$R_n(x) = xR_{n-1}(x) - \gamma_n R_{n-2}(x),$$

$$\gamma_{2k} = \frac{c}{1-c}(k + \beta - 1), \quad \gamma_{2k+1} = \frac{k}{1-c}, \quad k \geq 1. \qquad (3.16)$$

(B) *Meixner polynomials of the second kind.* For the polynomials which satisfy (3.1) under the conditions $\sigma^2 = 4g - d^2 > 0$, write

$$\delta = d/\sigma, \qquad \eta = 1 + h/g,$$

and choose $2f = \delta\eta\sigma$ (in contrast to the choice, $f = 0$, made in V-§4). If we set

$$M_n(x; \delta, \eta) = M_n(x) = \left(\frac{2}{\sigma} \right)^n P_n\left(\frac{x\sigma}{2} \right),$$

the recurrence formula becomes

$$M_{n+1}(x) = [x - (2n + \eta)\delta]M_n(x) - (\delta^2 + 1)n(n + \eta - 1)M_{n-1}(x),$$

$$n \geq 0. \qquad (3.17)$$

The recurrence formula (3.17) can be taken as the definition of the *Meixner polynomials of the second kind.* Meixner's generating function can now be written

$$[(1 + \delta t)^2 + t^2]^{-\eta/2}\exp\left\{ x \tan^{-1}\left(\frac{t}{1 + \delta t} \right) \right\} = \sum_{n=0}^{\infty} M_n(x)\frac{t^n}{n!}. \qquad (3.18)$$

Orthogonality in the positive-definite sense occurs for δ real and $\eta > 0$. The explicit orthogonality relation is

$$\int_{-\infty}^{\infty} M_k(x) M_n(x)w(x)\,dx = (\delta^2 + 1)^n n!\,(\eta)_n \int_{-\infty}^{\infty} w(x)\,dx\,\delta_{kn}, \qquad (3.19)$$

where $w(x) = w(x; \delta, \eta)$ is given by

$$w(x; \delta, \eta) = [\Gamma(\eta/2)]^{-2} \left| \Gamma\left(\frac{\eta + ix}{2}\right) \right|^2 \exp(-x \tan^{-1}\delta). \qquad (3.20)$$

When η is a positive integer, the weight function can be expressed in terms of elementary functions using known relations for the gamma function (Erdélyi [1, (1.2.8), (1.2.9)]):

$$w(x; \delta, 2N) = \frac{\pi x}{2 \sinh(\pi x/2)} \prod_{k=1}^{N-1} \left(1 + \frac{x^2}{4k^2}\right) \exp(-x \tan^{-1}\delta),$$

$$w(x; \delta, 2N + 1) = \mathrm{sech}(\pi x/2) \prod_{k=1}^{N} \left(1 + \frac{x^2}{(2k-1)^2}\right) \exp(-x \tan^{-1}\delta).$$

(An empty product, $\prod_{k=1}^{0} p_k$, is interpreted as 1.)

The Meixner polynomials of the second kind were later studied by F. Pollaczek [5] who considered

$$P_n^\lambda(x; \phi) = \frac{\sin^n \phi}{n!} M_n(2x; \delta, 2\lambda), \qquad \delta = \cot \phi, \qquad 0 < \phi < \pi. \qquad (3.21)$$

There is the hypergeometric representation

$$P_n^\lambda(x; \phi) = (n!)^{-1} (2\lambda)_n e^{in\phi} {}_2F_1(-n, \lambda + ix; 2; 1 - e^{-2i\phi}). \qquad (3.22)$$

For a formula analogous to the Rodrigues' type formulas due to L. Toscano, see Erdélyi [1, vol. 2]. See also (5.13) for Pollaczek's generalization.

When $\delta = 0$, we have symmetric polynomials so, as noted by Al-Salam (see V-§4), the related polynomials

$$U_n(x) = M_{2n}(x^{1/2}; 0; \eta), \qquad V_n(x) = x^{-1/2} M_{2n+1}(x^{1/2}; 0, \eta) \qquad (\eta > 0) \qquad (3.23)$$

form OPS with respect to the weight functions

$$u(x; \eta) = x^{-1/2} \left| \Gamma\left(\frac{\eta + ix^{1/2}}{2}\right) \right|^2, \qquad 0 < x < \infty,$$

and $v(x; \eta) = xu(x; \eta)$, respectively.

The corresponding recurrence formulas were given by Al-Salam:

$$U_{n+1}(x) = \{x - [8n^2 + (4n + 1)\eta]\}U_n(x)$$
$$- 2n(2n - 1)(2n + \eta - 2)(2n + \eta - 1)U_{n-1}(x)$$
$$V_{n+1}(x) = \{x - [2(2n + 1)^2 + (4n + 3)\eta]\}V_n(x)$$
$$- 2n(2n + 1)(2n + \eta - 1)(2n + \eta)V_{n-1}(x). \tag{3.24}$$

The particular case, $\delta = 0$, $\eta = k = 1, 2, 3, \ldots$, was studied by L. Carlitz [5] who considered

$$f_n^{(k)}(x) = (-i)^n M_n(ix; 0, k). \tag{3.25}$$

Carlitz obtains

$$\mathcal{L}[x^m f_n^{(k)}(x)] = n! \, (k)_n \delta_{mn}, \tag{3.26}$$

where the moments of $\mathcal{L}$ are the "Euler numbers of order k" defined by

$$\operatorname{sech}^k t = \sum_{m=0}^{\infty} E_m^{(k)} \frac{t^m}{m!}.$$

Carlitz also obtains the weight function for the real orthogonality relation corresponding to $i^n f_n^{(k)}(-ix) = M_n(x; 0, k)$ in the interesting form

$$\int_{-\infty}^{\infty} \cdots \int_{-\infty}^{\infty} \operatorname{sech} \pi t_1 \cdots \operatorname{sech} \pi t_{k-1}$$
$$\cdot \operatorname{sech} \pi(x - t_1 - \cdots - t_{k-1}) \, dt_1 \cdots dt_{k-1}.$$

The special case $\delta = 0$, $\eta = 1$ has also been considered by Hardy [1] and Weisner [1] while polynomials that correspond to the non-orthogonal case $\delta = \eta = 0$ have been encountered by H. A. Lauwerier [1].

As is the case with the Charlier polynomials, Carleman's criterian (1.14) can be applied to the recurrence formula (3.1) to verify that the Hamburger moment problems associated with the Meixner polynomials of both kinds are determined.

4 The Bessel Polynomials

The solution of the wave equation in spherical coordinates by the standard technique of separation of variables leads to Bessel's differential equation.

By a variation in this technique, Krall and Frink [1] obtain instead the differential equation

$$x^2 y'' + 2(x + 1)y' - n(n + 1)y = 0. \tag{4.1}$$

When n is a non-negative integer, (4.1) has a polynomial solution $y_n(x)$ which, with the standardization $y_n(0) = 1$, they name the *Bessel polynomial*. In the same paper, Krall and Frink introduce a generalization of (4.1):

$$x^2 y'' + (ax + b)y' - n(n + a - 1)y = 0,$$
$$b \neq 0, \quad a \neq 0, -1, -2, \ldots. \tag{4.2}$$

They define the *generalized Bessel polynomial* $y_n(x, a, b)$ as the polynomial solution of (4.2) satisfying the initial condition $y_n(0, a, b) = 1$.

It is easy to see that $y_n(bx, a, b)$ is independent of b. Thus it seems preferable to adopt the notation (W. A. Al-Salam [1])

$$Y_n^\alpha(x) = y_n(x, \alpha + 2, 2) \tag{4.3}$$

so that $Y_n^0(x) = y_n(x)$, the ordinary Bessel polynomial.

$Y_n^\alpha(x)$ satisfies the recurrence formula

$$2(n + \alpha + 1)(2n + \alpha)Y_{n+1}^\alpha(x)$$
$$= (2n + \alpha + 1)[(2n + \alpha)(2n + \alpha + 2)x + 2\alpha]Y_n^\alpha(x) \tag{4.4}$$
$$+ 2n(2n + \alpha + 2)Y_{n-1}^\alpha(x), \qquad n \geq 1,$$

$$Y_0^\alpha(x) = 1, \qquad Y_1^\alpha(x) = \frac{\alpha + 2}{2}x + 1.$$

Thus $\{Y_n^\alpha(x)\}$ is a quasi-definite OPS and Krall and Frink give the orthogonality relation

$$\frac{1}{2\pi i} \int_C Y_m^\alpha(z) Y_n^\alpha(z) \rho^\alpha(z) \, dz = \frac{2(-1)^{n+1} n!}{(2n + \alpha + 1)(\alpha + 1)_n} \delta_{mn} \tag{4.5}$$

where

$$\rho^\alpha(z) = \sum_{k=0}^\infty \frac{1}{(\alpha + 1)_k} \left(-\frac{2}{z} \right)^k \tag{4.6}$$

and the integration is around the unit circle.

They give the explicit representation

$$Y_n^\alpha(x) = \sum_{k=0}^{n} \binom{n}{k} (n + \alpha + 1)_k \left(\frac{x}{2}\right)^k. \tag{4.7}$$

From among the many other formulas given by these authors, we cite one of several connections between Bessel functions and Bessel polynomials:

$$y_n\left(\frac{1}{ir}\right) = \left(\frac{\pi r}{2}\right)^{1/2} e^{ir} [i^{-n-1} J_{n+1/2}(r) + i^n J_{-n-1/2}(r)]. \tag{4.8}$$

Burchnall [1] gave the generating function

$$(1 - 2xw))^{-1/2} \left(\frac{2}{1 + \sqrt{1 - 2xw}}\right)^\alpha \exp\left\{\frac{2w}{1 + \sqrt{1 - 2xw}}\right\}$$
$$= \sum_{n=0}^{\infty} Y_n^\alpha(x) \frac{w^n}{n!}. \tag{4.9}$$

Although Krall and Frink were the first to study the generalized Bessel polynomials in any great detail and their original paper was the starting point for the investigations of many writers, these polynomials occurred in the literature earlier. They were apparently first studied by Romanovsky [2] and were mentioned briefly by Bochner [1], Hahn [1], Hildebrandt [1] Krall [3], and encountered by Chaundy and Burchnall [1] in an investigation on differential equations. Romanovsky provided a number of identities subsequently rediscovered by others including the differentiation formula

$$2 D Y_n^\alpha(x) = n(n + \alpha + 1) Y_{n-1}^{\alpha+2}(x) \tag{4.10}$$

and the Rodrigues' type formula

$$Y_n^\alpha(x) = 2^{-n} x^{-\alpha} e^{2/x} D^n (x^{2n+\alpha} e^{-2/x}). \tag{4.11}$$

These, together with the differential equation (4.2), reflect the familial relationship of the Bessel polynomials with the classical orthogonal polynomials (V-§2(D)).

For numerous other formulas and further references, see the above mentioned papers and Rainville [1]. Rossum [3], [4] has studied the zeros of the Bessel polynomials and a general class of orthogonal polynomials which includes the Bessel polynomials. For an application of the Bessel polynomials to the inversion of Laplace transforms, see Salzer [1].

5 The Pollaczek Polynomials

The Pollaczek polynomials $P_n^\lambda(x) = P_n^\lambda(x; a, b)$ are defined by the recurrence formula

$$nP_n^\lambda(x) = 2[(n + \lambda + a - 1)x + b]P_{n-1}^\lambda(x) - (n + 2\lambda - 2)P_{n-2}^\lambda(x). \quad (5.1)$$

These polynomials, which reduce to the Ultraspherical polynomials when $a = b = 0$, were introduced by Pollaczek [1] for the case $\lambda = 1/2$ and extended to the general case by Szegö [7] (see Pollaczek [3]). They have the generating function

$$G(x, w) = (1 - we^{i\theta})^{-\lambda + it}(1 - we^{-i\theta})^{-\lambda - it} = \sum_{n=0}^{\infty} P_n^\lambda(x)w^n \quad (5.2)$$

where $t = (ax + b)(1 - x^2)^{-1/2}$, $x = \cos\theta$, $0 \leq \theta \leq \pi$. Written in terms of real functions, this becomes

$$G(x, w) = (1 - 2xw + w^2)^{-\lambda}\exp\left\{2t\,\tan^{-1}\frac{w(1 - x^2)^{1/2}}{1 - wx}\right\}. \quad (5.3)$$

The generating function (5.2) can be used to prove the orthogonality relation

$$\int_{-1}^{1} P_m^\lambda(x)P_n^\lambda(x)\rho^\lambda(x)\,dx = \frac{2\pi\Gamma(n + 2\lambda)}{2^{2\lambda}(n + \lambda + a)n!}\delta_{mn} \quad (5.4)$$

where

$$\rho^\lambda(x) = (\sin\theta)^{2\lambda - 1}e^{(2\theta - \pi)t}|\Gamma(\lambda + it)|^2$$
$$t = (ax + b)(1 - x^2)^{-1/2}, \qquad x = \cos\theta, \qquad 0 \leq \theta \leq \pi, \quad (5.5)$$

and the parameters satisfy the conditions $a \geq |b|$, $\lambda > 0$. (For different proofs, see Pollaczek [3] and Szegö [7].)

These polynomials can be represented by hypergeometric functions (Erdélyi [1]):

$$n!\,P_n^\lambda(x; a, b) = (2\lambda)_n e^{in\theta}\,{}_2F_1(-n, \lambda + it; 2\lambda; 1 - e^{-2i\theta}). \quad (5.6)$$

The generating function in the form (5.3) yields the formula

$$P_n^{\lambda + \lambda'}(x; a + a', b + b') = \sum_{k=0}^{n} P_k^\lambda(x; a, b)P_{n-k}^{\lambda'}(x; a', b'). \quad (5.7)$$

The Pollaczek ponomials are important as examples in the study of the asymptotic behavior of orthogonal polynomials. Since $\rho^\lambda(x)$ tends to zero exponentially as $x \to \pm 1$, $\log \rho^\lambda(x)$ is not integrable for $0 < \theta < \pi$. This means that the extensive asymptotic theory of Szegö [5] does not apply to these polynomials. Szegö [7] shows that the asymptotic behavior of $P_n^{1/2}(x; a, b)$ is markedly different from that of $P_n^{1/2}(x; 0, 0)$, the Legendre polynomials.

Thus these polynomials, which are amenable to asymptotic investigation, are important examples which may provide clues concerning the behavior of orthogonal polynomials when the weight function fails to satisfy certain integrability conditions. For additional studies of the asymptotic behavior of these polynomials, see Novikoff [1].

A generalization of the polynomials in (5.1) was provided by Pollaczek [6] by introducing a fourth parameter:

$$(n + c)P_n^\lambda(x) = 2[(n + \lambda + a + c - 1)x + b]P_{n-1}^\lambda(x)$$
$$- (n + 2\lambda + c - 2)P_{n-2}^\lambda(x), \tag{5.8}$$

$P_n^\lambda(x) = P_n^\lambda(x; a, b, c)$. In terms of the monic polynomials,

$$\hat{P}_n^\lambda(x) = \frac{(c + 1)_n}{2^n(\lambda + a + c)_n} P_n^\lambda(x),$$

the recurrence formula becomes

$$\hat{P}_n^\lambda(x) = \left[x + \frac{b}{n + \lambda + a + c - 1} \right] \hat{P}_{n-1}^\lambda(x)$$
$$- \frac{(n + c - 1)(n + 2\lambda + c - 2)}{4(n + \lambda + a + c - 1)(n + \lambda + a + c - 2)} \hat{P}_{n-2}^\lambda(x). \tag{5.9}$$

Pollaczek obtains a generating function which he uses to prove the orthogonality relation

$$\int_{-1}^1 P_m^\lambda(x) P_n^\lambda(x) \rho^\lambda(x)\,dx = \frac{2\pi[\Gamma(c + 1)]^2 \Gamma(n + 2\lambda + c)}{2^{2\lambda}(n + \lambda + a + c)\Gamma(n + c + 1)} \delta_{mn} \tag{5.10}$$

where

$$\rho^\lambda(x) = \rho^\lambda(x; a, b, c) = (\sin \theta)^{2\lambda - 1} e^{(2\theta - \pi)t} |\Gamma(\lambda + c + it)|^2$$
$$\cdot |_2F_1(1 - \lambda + it, c; c + \lambda + it; e^{2i\theta})|^{-2}, \tag{5.11}$$

and the parameters now satisfy either of the conditions
  (i) $a > |b|, 2\lambda + c > 0, c \geqq 0$
  (ii) $a > |b|, 2\lambda + c \geqq 1, c > -1$.
Note that the recurrence formula shows that $\{P_n{}^\lambda(x; a, b, c + 1)\}$ is the sequence of numerator polynomials for $\{P_n{}^\lambda(x; a, b, c)\}$.

Pollaczek notes that the limit

$$Q_n{}^\lambda(x; \phi, c) = \lim_{\epsilon \to 0} P_n{}^\lambda\left(\epsilon x + \cos \phi; \frac{\sin \phi}{\epsilon}, -\frac{\sin \phi \cos \phi}{\epsilon}, c\right)$$
$$= P_n{}^\lambda(\cos \phi; 0, x \sin \phi, c) \tag{5.12}$$

satisfies

$$(n + c)Q_n{}^\lambda(x) = 2[x \sin \phi + (n + \lambda + c - 1)\cos \phi]Q_{n-1}{}^\lambda(x)$$
$$- (n + 2\lambda + c - 2)Q_{n-2}{}^\lambda(x). \tag{5.13}$$

For these polynomials, he obtains the orthogonality relation

$$\int_{-\infty}^{\infty} Q_m{}^\lambda(x)Q_n{}^\lambda(x)\sigma^\lambda(x)\,dx = \frac{2\pi[\Gamma(c + 1)]^2 \Gamma(n + 2\lambda + c)}{(2 \sin \phi)^{2\lambda} \Gamma(n + c + 1)}\delta_{mn}, \tag{5.14}$$

where

$$\sigma^\lambda(x) = \sigma^\lambda(x; \phi, c) = e^{x(2\phi - \pi)}|\Gamma(\lambda + c + ix)|^2$$
$$\cdot |{}_2F_1(1 - \lambda + ix, c; c + \lambda + ix; e^{2i\phi})|^{-2} \tag{5.15}$$

and $0 < \phi < \pi$, with either $2\lambda + c > 0, c \geqq 0$ or $2\lambda + c \geqq 1, c > -1$.

Pollaczek [5] had earlier considered the special case $c = 0$ which gives essentially the Meixner polynomials of the second kind. For this case, he observes the limiting relationships with the Hermite and Laguerre polynomials:

$$H_n(x) = n! \lim_{\lambda \to \infty} \lambda^{-n/2} Q_n{}^\lambda\left(\frac{x\lambda^{1/2} - \lambda \cos \phi}{\sin \phi}; \phi, 0\right) \tag{5.16}$$

$$L_n^{(\alpha)}(x) = \lim_{\phi \to 0} Q_n{}^\lambda\left(\frac{-x + 2\lambda - 2\lambda \cos \phi}{2 \sin \phi}; \phi, 0\right), \qquad \alpha = 2\lambda - 1. \tag{5.17}$$

Yet another related OPS was obtained by Pollaczek [4]. These polynomials, $R_n(x) = R_n(x; a, b, c)$, satisfy the recurrence relation

$$n(n + a - 1)R_n(x) = [2(n + a)(n + a - 1)x + (n + a - 1)b - \gamma a]R_{n-1}(x)$$
$$- (n + a)(n + c - 1)R_{n-2}(x), \qquad (5.18)$$

where γ is either root of $a\gamma^2 + b\gamma + a - c = 0$.

For these polynomials, Pollaczek gives the orthogonality relation

$$\int_{-1}^{1} R_m(x)R_n(x)\tau(x)\,dx = \frac{\pi\Gamma(n + c + 1)}{2^{c+1}n!}\delta_{mn}, \qquad (5.19)$$

$$\tau(x) = \tau(x; a, b, c) = \frac{(\sin\theta)^{c+1}e^{(2\theta-\pi)t}}{1 - 2\gamma x + \gamma^2}|\Gamma(1 + \frac{c}{2} + it)|^2,$$

$$t = \frac{(2a - c)x + b}{2\sqrt{1 - x^2}}, \qquad x = \cos\theta, \qquad 0 \leq \theta \leq \pi. \qquad (5.20)$$

He also notes the relation with Jacobi polynomials:

$$R_n(x; a, 0, 2a) = \frac{\Gamma(a + 1/2)\Gamma(n + 2a)}{\Gamma(2a)\Gamma(n + a + 1/2)}P_n^{(a-\gamma/2, a+\gamma/2)}(x), \qquad \gamma = \pm 1. \qquad (5.21)$$

For a general theory of orthogonal polynomials whose recurrence formulas have coefficients which are rational functions of n and which includes the various Pollaczek polynomials, see Pollaczek [7]. For some related biorthogonal polynomials, see Pollaczek [2], [3].

6 Modified Lommel Polynomials

The Bessel function of order ν satisfies the well known recurrence

$$J_{\nu+1}(z) = 2\nu z^{-1}J_\nu(z) - J_{\nu-1}(z).$$

By iteration, this leads to the identity

$$J_{\nu+m}(z) = R_{m,\nu}(z)J_\nu(z) - R_{m-1,\nu+1}(z)J_{\nu-1}(z),$$

where $R_{m,\nu}(z)$ is a polynomial in z^{-1} and is called Lommel's polynomial (see Erdélyi [1, vol. 2, p. 34].)

It was noted by W. Hahn [3] that the polynomial

$$h_{n,\nu}(x) = R_{n,\nu}\left(\frac{1}{x}\right) \tag{6.1}$$

satisfies the recurrence formula

$$h_{n+1,\nu}(x) = 2x(n+\nu)h_{n,\nu}(x) - h_{n-1,\nu}(x) \tag{6.2}$$

so that $\{h_{n,\nu}(x)\}_{n=0}^{\infty}$ is an OPS (for $\nu \neq 0, -1, -2, \ldots$). The positive-definite case occurs if and only if $\nu > 0$ and the explicit orthogonality relation for the resulting *modified Lommel polynomials* was obtained by D. Dickinson [1].

Let $\{j_{\nu,n}\}_{n=-\infty}^{\infty}$ denote the nonzero zeros of $J_\nu(x)$ ordered by

$$\cdots < j_{\nu,-2} < j_{\nu,-1} < j_{\nu,0} < 0 < j_{\nu,1} < j_{\nu,2} < \cdots$$

Then Dickinson's relation is

$$\sum_{k=-\infty}^{\infty} h_{m,\nu}(x_{\nu,k})h_{m,\nu}(x_{\nu,k})x_{\nu,k}^2 = 2^{-n-1}(\nu)_{n+1}\delta_{mn}, \tag{6.3}$$

where

$$x_{\nu,k} = j_{\nu-1,k}^{-1}, \qquad k = 0, \pm 1, \pm 2, \ldots.$$

The true interval of orthogonality is thus $[j_{\nu,0}^{-1}, j_{\nu,1}^{-1}]$.

Dickinson gives the hypergeometric representation

$$h_{n,\nu}(x) = (\nu)_n(2x)^n {}_2F_3\left(-\frac{n}{2}, -\frac{n-1}{2}; \nu, -n, 1-\nu-n; -x^{-2}\right), \tag{6.4}$$

a number of (divergent) generating functions and the following relation with the ordinary Bessel polynomials, $y_n(x)$:

$$2h_{n,k+1/2}(x) = i^{-n}[y_{n+k}(ix)y_{k-1}(-ix) + (-1)^n y_{n+k}(-ix)y_{k-1}(ix)],$$

$$(y_{-1}(x) = 1) \qquad k = 0, 1, 2, \ldots \tag{6.5}$$

For $k = 0$ and $k = 1$, respectively, (6.5) yields

$$2h_{n,1/2}(x) = i^{-n}[y_n(ix) + (-1)^n y_n(-ix)]$$
$$2h_{n,3/2}(x) = i^{-n}[y_{n+1}(ix) + (-1)^n y_{n+1}(-ix)]. \tag{6.6}$$

This shows that the real parts of $y_n(ix)$ and the imaginary parts of $y_n(ix)$

SOME SPECIFIC SYSTEMS OF ORTHOGONAL POLYNOMIALS   189

both form positive-definite OPS. This was later observed independently by Al-Salam and Carlitz [1]. See also, W. A. Al-Salam [3]; also I-Ex. 4.7.

For the monic polynomials,

$$H_{n,\nu}(x) = 2^{-n}(\nu)_n^{-1} h_{n,\nu}(x), \qquad (6.7)$$

the corresponding recurrence formula is

$$H_{n+1,\nu}(x) = xH_{n,\nu}(x) - \frac{1}{4(n+\nu)(n+\nu-1)} H_{n-1,\nu}(x), \qquad n \geqq 0. \quad (6.8)$$

Thus $\{H_{n,\nu+1}(x)\}$ is the monic numerator OPS corresponding to $\{H_{n,\nu}(x)\}$. If one sets

$$H_{n,\nu}(x; c) = H_{n,\nu}(x) - cH_{n-1,\nu+1}(x), \qquad (6.9)$$

then $\{H_{n,\nu}(x; c)\}$ satisfies (6.8) for $n \geqq 1$ together with the initial conditions, $H_0(x; c) = 1$, $H_1(x; c) = x - c$ (cf. III-Ex. 4.2).

For real c, $\{H_{n,\nu}(x; c)\}$ is an OPS with respect to the distribution step function whose jumps occur at the points, $y_{\nu,k}^{-1}$, where $y_{\nu,k}$ are the nonzero solutions of

$$J_{\nu-1}(x) - cJ_\nu(x) = 0.$$

The corresponding jump, $A_{\nu,k}$, at $x = y_{\nu,k}^{-1}$ is given by

$$A_{\nu,k} = \operatorname*{Res}_{z=x_{\nu,k}} \frac{J_\nu(z)}{J_{\nu-1}(z) - cJ_\nu(z)}$$

(see Chihara [2]).

A re-examination of the recurrence formula (6.2) reveals that for fixed $x \neq 0$, $h_{n,\nu}(x)$ is a polynomial in ν. Writing

$$K_n(\nu) = K_n(\nu, x) = (2x)^{-n} h_{n,\nu}(x), \qquad (6.10)$$

the recurrence formula can be written

$$K_{n+1}(\nu) = (\nu + n)K_n(\nu) - (2x)^{-2}K_{n-1}(\nu), \qquad n \geqq 0. \quad (6.11)$$

Maki [2] has shown that for real $x \neq 0$, $\{K_n(\nu)\}$ satisfies the orthogonality relation

$$\sum_{i=1}^{\infty} K_m(\nu_i)K_n(\nu_i)A_i = (2x)^{-2n}\delta_{mn}, \qquad (6.12)$$

where the $\nu_i = \nu_i(x)$ are the poles and the $A_i = A_i(x)$ are the corresponding residues of the meromorphic function of ν,

$$F(\nu) = \frac{2xJ_{\nu+1}(1/x)}{J_\nu(1/x)}. \tag{6.13}$$

The true interval of orthogonality is $(-\infty, \nu_1)$ where $\nu_1 < 0$ and the Hamburger moment problem is determined.

Maki also notes that

$$K_n^{(s)}(\nu) = K_n(\nu + s) \tag{6.14}$$

so that (6.12) also yields the orthogonality relations for the numerator polynomials of order s.

7 Tricomi-Carlitz Polynomials

Tricomi [1] has studied the polynomials

$$t_n^{(\alpha)}(x) = \sum_{k=0}^{n} (-1)^k \binom{x-\alpha}{k} \frac{x^{n-k}}{(n-k)!} \tag{7.1}$$

which satisfy the recurrence

$$(n+1)t_{n+1}^{(\alpha)}(x) - (n+\alpha)t_n^{(\alpha)}(x) + xt_{n-1}^{(\alpha)}(x) = 0, \qquad n \geq 1, \tag{7.2}$$

$$t_0^{(\alpha)}(x) = 1, \qquad t_1^{(\alpha)}(x) = \alpha.$$

Tricomi noted that $\{t_n^{(\alpha)}(x)\}$ is not an OPS but Carlitz [3] observed that if one sets

$$f_n^{(\alpha)}(x) = x^n t_n^{(\alpha)}(x^{-2}), \tag{7.3}$$

then $\{f_n^{(\alpha)}(x)\}$ satisfies

$$(n+1)f_{n+1}^{(\alpha)}(x) - (n+\alpha)xf_n^{(\alpha)}(x) + f_{n-1}^{(\alpha)}(x) = 0, \qquad n \geq 1, \tag{7.4}$$

$$f_0^{(\alpha)}(x) = 1, \qquad f_1^{(\alpha)}(x) = \alpha x.$$

In terms of the corresponding monic polynomials,

$$F_n^{(\alpha)}(x) = \frac{n!}{(\alpha)_n} f_n^{(\alpha)}(x), \tag{7.5}$$

the recurrence is

$$F_{n+1}^{(\alpha)}(x) = xF_n^{(\alpha)}(x) - \frac{n}{(n+\alpha)(n+\alpha-1)}F_{n-1}^{(\alpha)}(x). \tag{7.6}$$

Thus $\{F_n^{(\alpha)}(x)\}$ is an OPS for $\alpha \neq 0, -1, -2, \ldots$ with the positive-definite case occuring for $\alpha > 0$. Carlitz proves that $\{f_n^{(\alpha)}(x)\}$ satisfies the orthogonality relation

$$\int_{-\infty}^{\infty} f_m^{(\alpha)}(x)f_n^{(\alpha)}(x)\,d\psi^{(\alpha)}(x) = \frac{2e^\alpha}{(n+\alpha)n!}\delta_{mn}, \tag{7.7}$$

where $\psi^{(\alpha)}$ is a step function whose jumps are

$$d\psi^{(\alpha)}(x) = \frac{(k+\alpha)^{k-1}e^{-k}}{k!} \quad \text{at } x = \pm(k+\alpha)^{-1/2}, \quad k = 0, 1, 2, \ldots$$

The generating function

$$\exp\left\{\frac{w}{x} - \frac{1-\alpha x^2}{x^2}\log(1-wx)\right\} = \sum_{n=0}^{\infty} f_n^{(\alpha)}(x)w^n \tag{7.8}$$

is readily verified from the recurrence (7.4).

8   OPS Related to Bernoulli Numbers

Stieltjes [1], [2] obtained the continued fraction expansion

$$\psi(z+h) - \psi(z+1-h) = \frac{2h-1|}{|z} + \frac{b_1|}{|z} + \frac{b_2|}{|z} + \cdots \tag{8.1}$$

where

$$\psi(z) = \frac{\Gamma'(z)}{\Gamma(z)},$$

$$b_n = \frac{n^2[n^2-(1-2h)^2]}{4(4n^2-1)}.$$

Stieltjes obtains the moments for the associated OPS in terms of Bernoulli polynomials.

Carlitz [5], generalizing earlier work by J. Touchard [1], rediscovered and extended Stieltjes' results by determining explicitly the OPS whose moments are $\beta_n(\lambda)$ defined by

$$\beta_n(\lambda) = \frac{B_{n+1}(\lambda) - B_{n+1}(0)}{(n+1)\lambda}, \tag{8.2}$$

where the $B_n(\lambda)$ are the Bernoulli polynomials:

$$\frac{xe^{\lambda x}}{e^x - 1} = \sum_{n=0}^{\infty} B_n(\lambda)\frac{x^n}{n!}.$$

(Stieltjes refers to $\lambda\beta_n(\lambda)$ as the "polynomes de Bernoulli.")
Carlitz defines

$$\Omega_n^{(\lambda)}(x) = \frac{(-1)^n(\lambda+1)_n n!}{2^n(1/2)_n} F_n^\lambda(1 - \lambda + 2x), \tag{8.3}$$

where $F_n^\lambda(x)$ is *Pasternak's polynomial*,

$$F_n^\lambda(x) = {}_3F_2[-n, n+1, (1+\lambda+x)/2; 1, \lambda+1; 1].$$

He then proves that

$$\mathcal{L}\{x^k\Omega_n^{(\lambda)}(x)\} = K_n^{(\lambda)}\delta_{mn} \tag{8.4}$$

where

$$\mathcal{L}[x^m] = \beta_m(\lambda),$$

$$K_n^{(\lambda)} = \frac{(-1)^n(n!)^2(1+\lambda)_n(1-\lambda)_n}{(2n+1)2^n[1 \cdot 3 \cdots (2n-1)]^2}.$$

Carlitz also obtains the orthogonality relation in the form

$$\int_{\alpha-i\infty}^{\alpha+i\infty} \Omega_m^{(\lambda)}(z)\Omega_n^{(\lambda)}(z)\frac{dz}{\sin(\pi z)\sin\pi(z-\lambda)} = \frac{2^{n+1}\lambda i}{\sin\pi\lambda}K_n^{(\lambda)}\delta_{mn}, \tag{8.5}$$

$$(-1 < \alpha < 0).$$

(See also, Al-Salam [4].) $\{\Omega_n^{(\lambda)}(x)\}$ is not a positive-definite OPS but the monic polynomials,

$$G_n^{(\lambda)}(x) = \frac{i^n n!(1+\lambda)_n}{2^n(1/2)_n} F_n(ix) = (-i)^n\Omega_n^{(\lambda)}\left(\frac{\lambda-1+ix}{2}\right), \tag{8.6}$$

satisfy the recurrence

$$G_{n+1}^{(\lambda)}(x) = xG_n^{(\lambda)}(x) - \frac{n^2(n^2 - \lambda^2)}{4n^2 - 1}G_{n-1}^{(\lambda)}(x) \qquad (8.7)$$

and hence is a positive-definite OPS for $-1 < \lambda < 1$. The corresponding orthogonality relation is

$$\int_{-\infty}^{\infty} G_m^{(\lambda)}(x)G_n^{(\lambda)}(x)\frac{dx}{\cos \pi\lambda + \cosh \pi x}$$
$$= \frac{2\lambda(1 - \lambda)_n(1 + \lambda)_n(n!)^2}{\sin \pi\lambda(2n + 1)[1 \cdot 3 \cdots (2n - 1)]^2}\delta_{mn}. \qquad (8.8)$$

$\Omega_n^{(\lambda)}(x)$ is related to the Hahn polynomials (V-(3.8)) by

$$\Omega_n^{(\lambda)}(x) = \frac{(-1)^n n!}{2^n(1/2)_n}p_n(-x - 1; 1, \lambda + 1, \lambda + 1). \qquad (8.9)$$

This thus provides a finite difference Rodrigues' type formula for $\Omega_n^{(\lambda)}(x)$. There is also a connection with Meixner polynomials given by Carlitz:

$$\Omega_n^{(1/2)}(x) = 2^{-n}(n!)^{-1}m_n(2x; 1, -1). \qquad (8.10)$$

For additional formulas for the case $\lambda = 0$, which was originally considered by Touchard, see Brafman [1] and Wyman and Moser [1].

Application of Carleman's criterian (1.11) to the recurrence formula (8.7) shows that the Hamburger moment problem associated with $\{G_n^{(\lambda)}(x)\}$ is a determined one.

9 OPS Related to Jacobi Elliptic Functions

Beginning with four J-fraction expansions given by Stieltjes [2], Carlitz [6] has studied four sets of orthogonal polynomials that can be defined by the following recurrence formulas.

$$A_{n+1}(x) = [x - (2n + 1)^2 a]A_n(x) - (2n - 1)(2n)^2(2n + 1)k^2 A_{n-1}(x) \qquad (9.1)$$

$$B_{n+1}(x) = [x - (2n + 2)^2 a]B_n(x) - 2n(2n + 1)^2(2n + 2)k^2 B_{n-1}(x) \qquad (9.2)$$

$$C_{n+1}(x) = xC_n(x) - \alpha_n C_{n-1}(x) \qquad (9.3)$$

$$D_{n+1}(x) = xD_n(x) - \beta_n D_{n-1}(x) \qquad (9.4)$$

where k is a parameter, $a = k^2 + 1$,

$$\alpha_{2m} = (2m)^2 k^2, \qquad \alpha_{2m+1} = (2m + 1)^2$$

$$\beta_{2m} = (2m)^2, \qquad \beta_{2m+1} = (2m + 1)^2 k^2.$$

If k is real, then all four sets are positive-definite OPS. Under the added assumption that $0 < k < 1$, Stieltjes obtained the distribution functions for the orthogonality relations satisfied by each of the four OPS. Working directly with the recurrences and independently of the theory of continued fractions, Carlitz also obtains these and gives the precise orthogonality relations:

$$\sum_{j=0}^{\infty} A_m(s_j^2) A_n(s_j^2) \frac{(2j + 1)q^{(2j+1)/2}}{1 - q^{2j+1}} = \frac{\pi^2}{kK^2}(2n)! \, (2n + 1)! \, k^{2n} \delta_{mn} \quad (9.5)$$

$$\sum_{j=1}^{\infty} B_m(t_j^2) B_n(t_j^2) \frac{j^3 q^j}{1 - q^{2j}} = \frac{2\pi^4}{k^2 K^4}(2n + 1)! \, (2n + 2)! \, k^{2n} \delta_{mn} \quad (9.6)$$

$$\sum_{j=-\infty}^{\infty} C_m(s_j) C_n(s_j) \frac{q^{(2j+1)/2}}{1 + q^{2j+1}} = \frac{kK}{\pi}(n!)^2 k^{2[n/2]} \delta_{mn} \quad (9.7)$$

$$\sum_{j=-\infty}^{\infty} D_m(t_j) D_n(t_j) \frac{q^j}{1 + q^{2j}} = \frac{K}{\pi}(n!)^2 k^{2[(n+1)/2]} \delta_{mn} \quad (9.8)$$

where

$$s_j = \frac{\pi}{2K}(2j + 1), \qquad t_j = \frac{\pi}{K}j, \qquad j = 0, \pm 1, \pm 2, \ldots$$

$$K = \int_0^{\infty} (1 - k^2 \sin^2 \varnothing)^{-1/2} \, d\varnothing,$$

and q is a certain constant related to k which appears in the theory of theta functions (in terms of which, the Jacobi elliptic functions can be defined). We will not attempt to describe the relation here (see Whittaker and Watson [1, pp. 478-479]-k is called the *modulus*) but mention that, for $0 < k < 1$, we have $0 < q < 1$.

Generating functions, which form the basis of his derivations of (9.5)–(9.8), together with a number of other relations are also given by Carlitz.

Carlitz also notes explicitly the orthogonality properties and the recurrence relations for the related OPS, $\{C_{2n}(x^{1/2})\}$ and $\{D_{2n}(x^{1/2})\}$. These relations can also be obtained for the corresponding KPS (with K-parameter 0), $\{x^{-1/2} C_{2n+1}(x^{1/2})\}$ and $\{x^{-1/2} D_{2n+1}(x^{1/2})\}$. Writing

$$F_n(x) = C_{2n}(x^{1/2}), \qquad F_n^*(x) = x^{-1/2}C_{2n+1}(x^{1/2}),$$
$$G_n(x) = D_{2n}(x^{1/2}), \qquad G_n^*(x) = x^{-1/2}D_{2n+1}(x^{1/2}). \tag{9.9}$$

the recurrence relations take the forms

$$F_{n+1}(x) = [x - (2n + 1)^2 - 4n^2k^2]F_n(x) - 4n^2(2n - 1)^2k^2F_{n-1}(x) \tag{9.10}$$

$$F_{n+1}^*(x) = [x - (2n + 1)^2 - (2n + 2)^2k^2]F_n^*(x) - 4n^2(2n + 1)^2k^2F_{n-1}^*(x) \tag{9.11}$$

$$G_{n+1}(x) = [x - (2n)^2 - (2n + 1)^2k^2]G_n(x) - 4n^2(2n - 1)^2k^2G_{n-1}(x) \tag{9.12}$$

$$G_{n+1}^*(x) = [x - (2n + 2)^2 - (2n + 1)^2k^2]G_n^*(x)$$
$$- 4n^2(2n + 1)^2k^2G_{n-1}^*(x). \tag{9.13}$$

The latter two were given by Al-Salam [5]. The explicit orthogonality relations can be obtained routinely from (9.7) and (9.8). The Hamburger moment problems associated with all of the above OPS are determined.

10   The q-Polynomials of Al-Salam and Carlitz

Let $q \neq 1$ be a fixed number and put

$$e(w) = e(q, w) = \sum_{k=0}^{\infty} \frac{w^k}{[q]_k} = \begin{cases} \prod_{m=0}^{\infty} (1 - q^m w)^{-1} & |q| < 1 \\ \prod_{n=1}^{\infty} (1 - q^{-n}w) & |q| > 1, \end{cases} \tag{10.1}$$

where $[b]_n$ is defined in (2.4).

Al-Salam and Carlitz [2] have defined two polynomials sets by means of the Brenke type generating functions

$$\frac{e(xw)}{e(w)e(aw)} = \sum_{n=0}^{\infty} U_n^{(a)}(x; q)\frac{w^n}{[q]_n} \tag{10.2}$$

$$\frac{e(w)e(aw)}{e(xw)} = \sum_{n=0}^{\infty} (-1)^n q^{n(n-1)/2} V_n^{(a)}(x; q)\frac{w^n}{[q]_n} \tag{10.3}$$

Using the identity

$$e(q^{-1}, w) = e(q, qw)^{-1},$$

it is readily verified that

$$V_n^{(a)}(x; q) = U_n^{(a)}(x; q^{-1}). \tag{10.4}$$

Writing

$$G(w) = \frac{e(xw)}{e(w)e(aw)},$$

one can verify the identity

$$(1 - w)(1 - aw)[G(w) - G(qw)] = w[x - (1 + a) + aw]G(w).$$

Using this identity, Al-Salam and Carlitz obtain the recurrence formula

$$U_{n+1}^{(a)}(x) = [x - (1 + a)q^n]U_n^{(a)}(x) + aq^{n-1}(1 - q^n)U_{n-1}^{(a)}(x), \qquad n \geq 0. \tag{10.5}$$

Because of (10.4), we also have

$$V_{n+1}^{(a)}(x) = [x - (1 + a)q^{-n}]V_n^{(a)}(x) - aq^{1-2n}(1 - q^n)V_{n-1}^{(a)}(x), \qquad n \geq 0. \tag{10.6}$$

Thus $\{U_n^{(a)}(x)\}$ is a positive-definite OPS for

$$a < 0, \qquad 0 < q < 1,$$

while $\{V_n^{(a)}(x)\}$ is a positive-definite OPS for

$$a > 0, \qquad 0 < q < 1.$$

For the former, Al-Salam and Caritz obtain the explicit orthogonality relation

$$\int_{-\infty}^{\infty} U_m^{(a)}(x)U_n^{(a)}(x) \, d\alpha^{(a)}(x) = C(-a)^n q^{n(n-1)/2}[q]_n \delta_{mn} \tag{10.7}$$

where

$$C = e(q)e(aq)e(a^{-1}q)\left\{ \sum_{n=0}^{\infty} \left[\frac{q^n}{e(q^{n+1}e(a^{-1}q^{n+1}))} - \frac{aq^n}{e(q^{n+1})e(aq^{n+1})} \right] \right\}$$

and $\alpha^{(a)}$ is a step function whose jumps occur at the points q^k and aq^k $(k = 0, 1, 2, \ldots)$. The jumps are

$$d\alpha^{(a)}(q^k) = e(aq)\frac{q^k}{[q]_k[a^{-1}q]_k}$$

$$d\alpha^{(a)}(aq^k) = -ae(a^{-1}q)\frac{q^k}{[q]_k[aq]_k}$$

$$k = 0, 1, 2, \ldots$$

They also identify the corresponding moments:

$$\int_{-\infty}^{\infty} x^n \, d\alpha^{(a)}(x) = C \sum_{k=0}^{n} \begin{bmatrix} n \\ k \end{bmatrix} a^k \qquad (10.8)$$

where $\begin{bmatrix} n \\ k \end{bmatrix}$ is given by (2.5).

For $\{V_n^{(a)}(x)\}$ they obtain

$$\int_{-\infty}^{\infty} V_m^{(a)}(x)V_n^{(a)}(x) \, d\beta^{(a)}(x) = Ka^n q^{-n^2}[q]_n \delta_{mn} \qquad (10.9)$$

where

$$K = \sum_{n=0}^{\infty} \frac{a^n q^{n^2}}{[q]_n[aq]_n},$$

and the spectrum of $\beta^{(a)}$ is $\{1, q^{-1}, q^{-2}, \ldots\}$ with

$$d\beta^{(a)}(q^{-k}) = \frac{a^k q^{k^2}}{[q]_k[aq]_k}, \qquad k = 0, 1, 2, \ldots$$

However, this provides a non-decreasing function only for $0 < aq < 1$.

The corresponding moments are identified as

$$\int_{-\infty}^{\infty} x^n \, d\beta^{(a)}(x) = K \sum_{k=0}^{n} \begin{bmatrix} n \\ k \end{bmatrix} a^k q^{k(k-n)}. \qquad (10.10)$$

In the case of $\{U_n^{(a)}(x)\}$ it is clear that the Hamburger moment problem is determined and the true interval of orthogonality is $[a, 1]$. For $\{V_n^{(a)}(x)\}$ the true interval of orthogonality is a subset of $[1, \infty]$. The true interval of orthogonality is precisely $[1, \infty]$ if and only if either

$$0 < a \leqq 1 \quad \text{or} \quad q^{-1} \leqq a \qquad (0 < q < 1).$$

Moreover, the corresponding Hamburger moment problem is determined if and only if

$$0 < a \leqq q \quad \text{or} \quad q^{-1} \leqq a$$

(see Chihara [8]). In particular, this means that when $1 < a < q^{-1}$, the true interval of orthogonality is of the form $[c, \infty]$ with $c > 1$.

11 Wall Polynomials

The *Wall polynomials*, $W_n(x) = W_n(x; b, q)$, will be defined directly by the recurrence formula,

$$W_{n+1}(x) = \{x - [b + q - (1 + q)bq^n]q^n\}W_n(x)$$
$$- b(1 - q^n)(1 - bq^{n-1})q^{2n}W_{n-1}(x), \qquad n \geqq 0. \tag{11.1}$$

These polynomials occur as the partial denominators of a certain J-fraction expansion that can be obtained by "contraction" from an S-fraction obtained by Wall [1] in a number theoretic study. Wall's work requires that $0 < b < 1$ and $0 < q < 1$ but for the study of (11.1) it is only necessary to assume $b \neq 1$, $bq \neq 0$, $q^k \neq 1$, b^{-1} $(k \geqq 1)$.

The Wall polynomials have the explicit representation (Chihara [6])

$$W_n(x; b, q) = (-1)^n[b]_n q^{n(n+1)/2} \sum_{k=0}^{n} \begin{bmatrix} n \\ k \end{bmatrix} \frac{q^{k(k-1)/2}(-q^{-n}x)^k}{[b]_k}. \tag{11.2}$$

(For the notation, see (2.4), (2.5).)

They form an OPS with respect to the formal moment sequence given by

$$\mu_n(b, q) = [b]_n q^n, \qquad n = 0, 1, 2, \ldots \tag{11.3}$$

The identity (10.1) shows that

$$\mu_n(b, q) = \frac{e(q, bq^n)q^n}{e(q, b)} = e(q, b)^{-1} \sum_{k=0}^{\infty} \frac{q^{n(k+1)}b^k}{[q]_k}, \tag{11.4}$$

at least when the series converges ($|q| > 1$ or $0 < |b| < 1$, $0 < |q| < 1$). Then the Wall polynomials will be orthogonal with respect to a step function ψ whose jumps are

$$d\psi(q^{k+1}) = d\psi(q^{k+1}; b, q) = \frac{b^k}{e(q, b)[q]_k}, \qquad k = 0, 1, 2, \ldots \tag{11.5}$$

This distribution function follows, for $0 < q < 1$ and $0 < b < 1$, from Wall's continued fraction by applying the Stieltjes inversion formula. It was obtained by Geronimus [4] without reference to the continued fraction under the implicit conditions $q > 1$ and $b < 0$. The corresponding orthogonality relation is

$$\int_{-\infty}^{\infty} W_m(x)W_n(x)\,d\psi(x) = [b]_n[q]_n b^n q^{n(n+1)}\delta_{mn}. \tag{11.6}$$

It is quickly seen from (11.1) that orthogonality in the positive-definite sense occurs in the following cases:

      (A) $0 < b < 1,$ $0 < q < 1$

      (B) $b < 0,$ $q > 1$

      (C) $b > 1,$ $q > 1.$

In cases (A) and (B), the function ψ above is non-decreasing. In case (C), (11.6) remains true but ψ is not non-decreasing. However, in this case, orthogonality relations in the positive-definite sense can be obtained by observing that the Wall polynomials can be reduced to generalized Stieltjes-Wigert polynomials:

$$W_n(x; b, q) = (-1)^n b^n q^{-n/2} S_n(-b^{-1}q^{1/2}x; b^{-1}, q^{-1}). \tag{11.7}$$

The Wall polynomials have the Brenke type generating function

$$A(w)B(xw) = \sum_{n=0}^{\infty} W_n(x; b, q)\frac{w^n}{[b]_n[q]_n} \tag{11.8}$$

where

$$A(w) = \sum_{n=0}^{\infty} \frac{(-1)^n q^{n(n+1)/2}w^n}{[q]_n} = \prod_{n=1}^{\infty} (1 - q^n w)$$

$$B(w) = \sum_{n=0}^{\infty} \frac{w^n}{[b]_n[q]_n}.$$

For further details, see Chihara [6].

An examination of the moments (11.3) shows that $\{W_n(x; bq, q)\}$ is the monic KPS with K-parameter 0 which corresponds to $\{W_n(x; b, q)\}$. It follows that the polynomials

$$Y_{2m}(x) = W_m(x^2; b, q), \qquad Y_{2m+1}(x) = xW_m(x^2; bq, q) \tag{11.9}$$

satisfy

$$\int_{-\infty}^{\infty} Y_m(x) Y_n(x) (\operatorname{sgn} x) \, d\psi(x^2) = K_n \delta_{mn} \tag{11.10}$$

where $K_{2m} = b^m q^{m(m+1)} [b]_m [q]_m$, $K_{2m+1} = b^m q^{(m+1)^2} [b]_{m+1} [q]_m$.
The corresponding recurrence formula is

$$Y_n(x) = x Y_{n-1}(x) - \gamma_n Y_{n-2}(x), \qquad n \geqq 1,$$
$$\gamma_{2k} = (1 - bq^{k-1}) q^k, \qquad \gamma_{2k+1} = b(1 - q^k) q^k, \qquad k \geqq 1. \tag{11.11}$$

A Brenke type generating function can be written using (11.8).

The polynomials $Y_n(x)$ are the numerator polynomials for the polynomials $X_n(x)$ satisfying

$$X_n(x) = x X_{n-1}(x) - \gamma_{n-1} X_{n-2}(x), \qquad \gamma_1 = b, n \geqq 1. \tag{11.12}$$

J. Goldberg [1] has shown that if $0 < b < q$, then the distribution function $\phi^{(k)}$ corresponding to $\{X_n^{(k)}(x)\}$ has a positive jump at the origin for even values of $k \geqq 0$ while $\phi^{(k)}$ is continuous at 0 for odd k. If $0 < q < b < 1$, then $\phi^{(k)}$ is continuous at 0 for every value of k.

The Hamburger moment problem corresponding to $\{W_n(x; b, q)\}$ is determined in the positive-definite case (A) $(0 < b < 1, 0 < q < 1)$ since the true interval of orthogonality is bounded in this case. In case (B) $(b < 0, q > 1)$ and case (C) $(b > 1, q > 1)$, the moment problems are indeterminate (see Chihara [5, Theorem 4.3]). In case (C), this conclusion also follows from (11.7).

Still another OPS closely related to the Wall polynomials is the set $\{Z_n(x; b, q)\}$ defined by

$$Z_{2n}(x) = W_n(x^2; b, q) + (1 - q^n) x W_{n-1}(x^2; bq, q)$$
$$Z_{2n+1}(x) = x W_n(x^2; bq, q) + (1 - b) W_n(x^2; b, q). \tag{11.13}$$

They satisfy the orthogonality relation

$$\int_{-\infty}^{\infty} Z_m(x) Z_n(x) \, d\psi(x) = K_n \delta_{mn} \tag{11.14}$$

where

$$K_{2n} = b^n q^{n^2} [b]_n [q]_n, \qquad K_{2n+1} = b^{n+1} q^{n(n+1)} [b]_{n+1} [q]_n,$$

and ψ is the step function which has the jump

$$\frac{\alpha(1 \mp q^{k/2})b^k}{2[q]_k}, \qquad \alpha = \begin{cases} \prod\limits_{k=0}^{\infty} (1 - bq^k) & \text{for } |q| < 1 \\ \prod\limits_{m=1}^{\infty} (1 - bq^{-m})^{-1} & \text{for } |q| > 1 \end{cases}$$

at the point $\pm q^{k/2}$ ($k = 0, 1, 2, \ldots$). Here, b and q must be restricted by $b \neq 0$, $q \neq 0$, 1, and $bq^m \neq 1$ ($m = 0, 1, 2, \ldots$) with the positive-definite case occuring for $0 < q < 1$, $0 < b < 1$ (Chihara [11]).

The recurrence relation is

$$\begin{aligned} Z_{2m+1}(x) &= [x + (1 - b)q^m]Z_{2m}(x) - (1 - q^m)q^m Z_{2m-1}(x) \\ Z_{2m+2}(x) &= [x + (b - q)q^m]Z_{2m+1}(x) - b(1 - bq^m)q^m Z_{2m}(x) \end{aligned} \qquad (11.15)$$

and they have the Brenke type generating function

$$A(w)B(xw) = \sum_{n=0}^{\infty} b_n Z_n(x)w^n \qquad (11.16)$$

where

$$A(w) = (1 + w) \sum_{k=0}^{\infty} \frac{(-1)^k q^{k(k+1)/2} w^{2k}}{[q]_k},$$

$$B(w) = \sum_{k=0}^{\infty} b_k w^k,$$

$$b_{2m} = \{[b]_m[q]_m\}^{-1}, \qquad b_{2m+1} = \{(1 - b)[bq]_m[q]_m\}^{-1}.$$

12 Associated Legendre Polynomials

The associated Legendre polynomials, $P_n(\nu, x)$, are defined by the recurrence formula

$$(n + \nu)P_n(\nu, x) = (2n + 2\nu - 1)xP_{n-1}(\nu, x) - (n + \nu - 1)P_{n-2}(\nu, x). \quad (12.1)$$

They are clearly related to the ordinary Legendre polynomials, $P_n(x)$, by

$$P_n(0, x) = P_n(x), \qquad P_n(k, x) = P_n^{(k)}(x), \qquad k = 0, 1, 2, \ldots \quad (12.2)$$

where $P_n^{(k)}(x)$ denotes the numerator polynomial of order k.

$\{P_n(\nu, x)\}$ is an OPS for 2ν not a negative integer and the positive-definite case occurs for $\nu > -1/2$. For the case $\nu > 0$, Barrucand and Dickinson [2] have obtained the explicit orthogonality relation

$$\int_{-1}^{1} P_m(\nu, x) P_n(\nu, x) w(\nu, x)\, dx = \frac{2}{2n + 2\nu - 1} \delta_{mn}$$

$$w(\nu, x) = \left\{ [\nu Q_{\nu-1}(x)]^2 + \left[\frac{\nu\pi}{2} P_{\nu-1}(x) \right]^2 \right\}^{-1}, \qquad \nu > 0. \tag{12.3}$$

Here $Q_\nu(x)$ denotes the Legendre function of the second kind (see Erdélyi [1, vol. 1]). In the limiting case $\nu = 0$, (12.3) reduces to the usual orthogonality relation for the Legendre polynomials with the aid of the limit

$$\lim_{\nu \to 0} \nu Q_{\nu-1}(x) = 1.$$

In the special case $\nu = 1$, the weight function was obtained by Sherman [1]:

$$w(1, x) = 4\left\{ \left[\log\left(\frac{1 + x}{1 - x} \right) \right]^2 + \pi^2 \right\}^{-1}. \tag{12.4}$$

Barrucand and Dickinson derive the generating function

$$\frac{\nu}{w^\nu(1 - 2xw + w^2)^{1/2}} \int_0^w \frac{t^{\nu-1}}{(1 - 2xw + w^2)^{1/2}}\, dt = \sum_{n=0}^\infty P_n(\nu, x) w^n, \tag{12.5}$$

$$\nu \neq 0.$$

Expressing the left side of (12.5) in terms of the generating function for Legendre polynomials, they then deduce

$$P_n(\nu, x) = \nu \sum_{k=0}^n \frac{P_k(x) P_{n-k}(x)}{k + \nu}, \qquad \nu \neq 0, \tag{12.6}$$

together with the limiting case

$$P_n(\infty, x) = U_n(x) = \sum_{k=0}^n P_k(x) P_{n-k}(x). \tag{12.7}$$

The case $\nu = 1$ is a formula of Christoffel (see Hobson [1 Ch. 2]).
For the corresponding monic polynomials,

$$\hat{P}_n(\nu, x) = \frac{2^n (\nu + 1/2)_n}{(\nu + 1)_n} P_n(\nu, x), \tag{12.8}$$

the recurrence formula (12.1) becomes

$$\hat{P}_{n+1}(\nu, x) = x\hat{P}_n(\nu, x) - \frac{(n + \nu)^2}{4(n + \nu)^2 - 1} \hat{P}_{n-1}(\nu, x), \qquad n \geq 0. \quad (12.9)$$

The recurrence formula and orthogonality relations for the related OPS, $\{P_{2m}(\nu, x^{1/2})\}$ and $\{x^{-1/2} P_{2m+1}(\nu, x^{1/2})\}$ can then be written immediately.

For additional properties of the associated Legendre polynomials, see in addition to the papers already quoted, Humbert [1], Palama [1], and Dickinson [2] where the orthogonality relation is expressed in terms of a complex integral. For a connection between the polynomials for $\nu = 1/2$ and $\nu = 3/2$ and the polynomials orthogonal with respect to the weight function, $[(1 - x^2)(1 - k^2 x^2)]^{-1/2}, -1 < x < 1$, see Rees [1].

A related OPS was studied by R. H. Boyer [1] in connection with a discrete analogue of Laplace's equation with axial symmetry. Boyer considers a partial difference equation which leads after separation of variables to the difference equation

$$(2n + 1)p_{n+1}(x) = 4n(1 - 2x)p_n(x) - (2n - 1)p_{n-1}(x), \qquad n \geq 1, \quad (12.10)$$

$$p_0(x) = 1, \qquad p_1(x) = 1 - x.$$

Upon writing

$$B_n(x) = p_n\left(\frac{1 - x}{2}\right),$$

(12.10) becomes

$$(2n - 1)B_n(x) = 4(n - 1)xB_{n-1}(x) - (2n - 3)B_{n-2}(x), \qquad n \geq 2, \quad (12.11)$$

$$B_0(x) = 1, \qquad B_1(x) = \frac{1}{2}(x + 1).$$

From this it is evident that $\{P_n(1/2, x)\}$ is the numerator OPS corresponding to $\{B_n(x)\}$.

Boyer does not observe that $\{p_n(x)\}$ is an OPS but he obtains two biorthogonality relations involving these polynomials and gives an application to a boundary value problem.

13   Miscellaneous OPS

We mention rather briefly some further examples of particular systems of orthogonal polynomials that have appeared in the literature. In many cases, relatively little is known about the polynomials themselves.

(A) Heine [1, vol. 1, pp. 294-296] showed that the orthogonal polynomials, $P_n(x)$, corresponding to the weight function

$$w(x) = [\psi(x)]^{-1/2}, \qquad 0 < x < \alpha, \tag{13.1}$$

where $\psi(x) = x(\alpha - x)(\beta - x)$, $0 < \alpha < \beta$, satisfy second order differential equations of the form

$$2\psi(x)(x - \gamma)y'' + [(x - \gamma)\psi'(x) - 2\psi(x)]y' + [a + bx - n(2n - 1)x^2]y$$
$$= 0, \qquad y = P_n(x), \tag{13.2}$$

Here a, b and γ are certain constants not given explicitly. These orthogonal polynomials are related to the Jacobi elliptic functions.

(B) Szegö [2] and Bernstein [1] studied the orthogonal polynomials corresponding to weight functions of each of the three forms,

$$
\begin{aligned}
w_1(x) &= (1 - x^2)^{-1/2}[\rho(x)]^{-1} \\
w_2(x) &= (1 - x^2)^{1/2}[\rho(x)]^{-1} \qquad -1 < x < 1, \qquad (13.3)\\
w_3(x) &= \left(\frac{1 - x}{1 + x}\right)^{1/2}[\rho(x)]^{-1}
\end{aligned}
$$

$\rho(x)$ being a polynomial.

In all three cases, explicit formulas are obtained for all but a finite number of the corresponding orthogonal polynomials (see Szegö [5]).

(C) For the orthogonal polynomials satisfying the Tchebichef recurrence relation

$$P_n(x) = 2xP_{n-1}(x) - P_{n-2}(x)$$

but with the general initial conditions

$$P_0(x) = 1, \qquad P_1(x) = ax - b, \qquad a \neq 0,$$

Geronimus [1] obtained the representation

$$P_n(x) = aT_n(x) + (a - 1)U_{n-2}(x) - bU_{n-1}(x). \tag{13.4}$$

(i) For the special case $b = 0$, Geronimus showed that the polynomials are orthogonal over $(-1, 1)$ with respect to the weight function

$$w(x) = \frac{(1 - x^2)^{1/2}}{1 - \mu x^2}, \qquad \mu = 2a - a^2 \leqq 1. \tag{13.5}$$

A second order differential equation is also given.

When $b \neq 0$, explicit orthogonality relations can be obtained for two values of a (Chihara [2]).

(ii) For $a = 1$, the distribution function ψ_1 is given by

$$\psi_1(x) = \frac{1}{\pi} \int_{-1}^{x} \frac{(1 - t^2)^{1/2}}{1 + b^2 - t} dt \qquad \text{for } -1 \leqq x \leqq 1 \qquad (13.6)$$

and for $x \notin [-1, 1]$, ψ_1 is a step function with a single jump of magnitude $|b|(1 + b^2)^{1/2}$ at the point $(\text{sgn } b)(1 + b^2)^{1/2}$.

(iii) For $a = 2$, the corresponding distribution function ψ_2 is given by

$$\psi_2(x) = \frac{2}{\pi} \int_{-1}^{x} \frac{(1 - t^2)^{1/2} dt}{1 + b^2 - 2bt} \qquad \text{for } -1 \leqq x \leqq 1 \qquad (13.7)$$

while ψ_2 is constant on $(-\infty, -1]$ and on $[1, \infty)$ except when $|b| > 1$ in which case ψ_2 has a single jump of magnitude $1 - b^{-2}$ at the point $(b/2) + (2b)^{-1}$.

(iv) The weight function

$$v(x) = \frac{1}{(1 - \mu x^2)(1 - x^2)^{1/2}}, \qquad -1 < x < 1, \qquad \mu < 1, \quad (13.8)$$

can be considered a companion to (13.5). Griñspun [1] obtained the corresponding orthogonal polynomials in the form

$$P_n(x) = T_n(x) + \frac{\alpha - 1}{\alpha + 1} T_{n-2}(x), \qquad \alpha = \sqrt{1 - \mu}. \qquad (13.9)$$

He also gives a number of other relations including the three term recurrence relation (see below), a generating function and a second order differential equation with polynomial coefficients.

For comparison purposes, the recurrence relations for the monic polynomials, $\hat{P}_n(x)$, corresponding to the above four cases are given.

$$\hat{P}_n(x) = x\hat{P}_{n-1}(x) - \lambda_n \hat{P}_{n-2}(x), \qquad n \geqq 2, \qquad (13.10)$$

where

$$\hat{P}_n(x) = 2^{1-n}[T_n(x) + (1 - \frac{1}{a})U_{n-2}(x)]$$

$$\lambda_2 = \frac{1}{2a}, \qquad \lambda_n = \frac{1}{4} \qquad \text{for } n \geqq 3; \qquad \text{(i)}$$

$$\hat{P}_n(x) = 2^{1-n}[T_n(x) - bU_{n-1}(x)]$$

$$\lambda_2 = \frac{1}{2}, \qquad \lambda_n = \frac{1}{4} \qquad \text{for } n \geqq 3 \quad (\text{note that } \hat{P}_1(x) = x - b); \qquad \text{(ii)}$$

$$\hat{P}_n(x) = 2^{-n}[U_n(x) - bU_{n-1}(x)]$$

$$\lambda_n = \frac{1}{4} \quad \text{for } n \geqq 2 \quad (\hat{P}_1(x) = x - b/2);$$

(iii)

$$\hat{P}_n(x) = 2^{1-n}\left[T_n(x) + \frac{\alpha - 1}{\alpha + 1}T_{n-2}(x)\right]$$

$$\lambda_2 = \frac{1}{1 + \alpha}, \quad \lambda_3 = \frac{\alpha}{2(1 + \alpha)}, \quad \lambda_n = \frac{1}{4} \quad \text{for } n \geqq 4.$$

(iv)

(D) Sherman [1] studied the numerator polynomials corresponding to the classical orthogonal polynomials. In particular, he showed that if m and n are non-negative integers with $m + n \geqq 1$, then the numerator polynomials corresponding to the Jacobi polynomials, $P_n^{(m,n)}(x)$, are orthogonal with respect to a weight function of the form

$$w(x) = \frac{(1 + x)^m(1 - x)^n}{\left[Q(x) + (1 + x)^m(1 - x)^n\log\left(\frac{1 + x}{1 - x}\right)\right]^2 + (1 + x)^{2m}(1 - x)^{2n}\pi^2}$$

$$-1 < x < 1,$$

(13.11)

where $Q(x)$ is a polynomial of degree $\leqq m + n - 1$. (For the case $m = n = 0$, $w(x)$ is given explicitly—see (12.4).)

(E) Ahiezer [1] studied the orthogonal polynomials corresponding to the weight function

$$w(x)$$

$$= \begin{cases} |c - x|\{(1 - x^2)a - x)(b - x)\}^{-1/2} & -1 < x < a, b < x < 1, \\ 0 & \text{otherwise .} \end{cases}$$

(13.12)

Here $-1 < a < b < 1$ and c depends on a and b. These polynomials are related to elliptic functions.

Later, Ahiezer [2] studied the asymptotic properties of certain orthogonal polynomials whose weight functions vanish outside the union of a finite number of disjoint, bounded intervals.

(F) Krall [2] showed that the orthogonal polynomial, $P_n(x)$, corresponding to the distribution function

$$\psi(x) = \begin{cases} -(1 + a) & x \leqq -1 \\ ax & -1 < x < 1 \\ 1 + a & x \geqq 1 \end{cases}$$

(13.13)

satisfies the fourth order differential equation

$$(x^2 - 1)^2 y^{iv} + 8x(x^2 - 1)y''' + (4a + 12)(x^2 - 1)y'' + 8axy'$$
$$= n(n + 1)(n^2 + n + 4a - 2)y, \qquad y = P_n(x). \tag{13.14}$$

A second order differential equation was obtained by Shore [1].

Later Krall (see Krall and Sheffer [1]) obtained for the polynomials orthogonal with respect to the distribution function

$$\phi(x) = \begin{cases} 1 + 2(a - 2)^{-1} & x \geqq 0 \\ e^x & -\infty < x < 0 \end{cases} \tag{13.15}$$

the differential equation

$$x^2 y^{iv} + (2x^2 + 4x)y''' + [x^2 + (a + 4)x]y'' + (ax + a - 2)y'$$
$$= [an + n(n - 1)]y, \qquad y = P_n(x). \tag{13.16}$$

(G) Rees [1] studied the orthogonal polynomials corresponding to the weight function

$$w(x) = [(1 - x^2)(1 - k^2 x^2)]^{-1/2}, \qquad -1 < x < 1. \tag{13.17}$$

A second order differential equation satisfied by the orthogonal polynomial of degree n is given. The coefficients of this differential equation are polynomials which depend on n and involve a parameter for which a recurrence formula is given. There is a connection between these polynomials and certain associated Legendre polynomials (§13).

(H) Hammer and Wicke [1] have obtained a quadrature formula which involves the zeros of the orthogonal polynomials corresponding to the two weight functions

$$w_1(x) = (1 - x^{1/2})^k, \qquad w_2(x) = x^{-1/2}(1 - x^{1/2})^k, \qquad 0 < x < 1. \tag{13.18}$$

No general formulas for the corresponding orthogonal polynomials are known.

(I) Konoplev [1] has studied the asymptotic properties of the orthonormal polynomials associated with the weight function

$$w(x) = (1 - x)^{\alpha}(1 + x)^{\beta}|x - x_0|^{\gamma}, \qquad -1 < x < 1,$$
$$\alpha > -1, \qquad \beta > -1, \qquad \gamma \geqq 0, \qquad |x_0| < 1. \tag{13.19}$$

He obtains a differential-recurrence formula for the polynomials. See also Konoplev [2] where weight functions leading to irregular aymptotic behavior are discussed.

(J) Barrucand and Dickinson [1] have shown that the OPS defined by

$$R_n(x) = xR_{n-1}(x) - \gamma_n R_{n-2}(x),$$

$$\gamma_{3m-1} = \frac{m}{2m-1}, \qquad \gamma_{3m} = 1, \qquad \gamma_{3m+1} = \frac{m}{2m+1}, \qquad m \geq 1, \tag{13.20}$$

are orthogonal with respect to a distribution function whose spectrum is $[-b, -1] \cup [-a, a] \cup [1, b]$, $2a = -1 + 5^{1/2}$, $2b = 1 + 5^{1/2}$. These polynomials can be expressed in terms of Legendre polynomials:

$$R_{3m}(x) = 2^{-n}[(1/2)_n]^{-1} n! \, P_n(x^3 - 2x),$$

$$R_{3m+1}(x) = 2^{-n}[(3/2)_n]^{-1}(n+1)! \, (x^2 - 1)^{-1}$$
$$\cdot [\, P_{n+1}(x^3 - 2x) + xP_n(x^3 - 2x)] \tag{13.21}$$

$$R_{3m+2}(x) = 2^{-n}[(3/2)_n]^{-1}(n+1)! \, (x^2 - 1)^{-1}$$
$$\cdot [xP_{n+1}(x^3 - 2x) + P_n(x^3 - 2x)].$$

(K) An eigenvalue problem in neutron-transport theory leads to an expansion of the eigensolutions in terms of Legendre polynomials. The coefficients in this expansion are polynomials which can be defined by the recurrence formula

$$(n+1)h_{n+1}(x) = (2n+1)(1 - cf_n)xh_n(x) - nh_{n-1}(x)$$

$$0 < c < 1, \qquad f_0 = 1, \qquad |f_n| < 1 \qquad \text{for } n > 0, \tag{13.22}$$

$$\sum_{n=0}^{\infty} (2n+1)|f_n| < \infty.$$

The parameters c and f_n depend upon the particular scattering law involved in the neutron-transport problem. Inönü [1] has noted the orthogonality of these polynomials and shown that they satisfy an orthogonality relation of the form

$$\int_{-v_1}^{v_1} h_m(x)h_n(x)\,d\omega(x) = \int_{-1}^{1} h_m(x)h_n(x)\rho(x)\,dx$$
$$+ \sum_{i=1}^{M} [h_m(v_i)h_n(v_i) + h_m(-v_i)h_n(-v_i)]J_i = K_n\delta_{mn}. \tag{13.23}$$

The even weight function, ρ, the normalization constant, K_n, the spectral points and jumps, v_i and J_i, all depend upon c and $\{f_n\}$ and Inönü gives formulas for them. The number, $2M$, of discrete spectral points all of which lie outside $[-1, 1]$ also depends on these parameters but is always finite.

NOTES

CHAPTER I. Most of the material of this chapter belongs to the classical lore of the subject. The standard reference is the treatise of Szegö [5] to which we defer for original sources not given here. A very readable introduction is Jackson [1]. The recent book by Freud [1] is highly recommended, especially for material not treated in this book. The older monograph of Shohat [1] is still interesting and valuable. Other good and fairly comprehensive treatments can be found in Sansone [1] and Natanson [1]. Shohat, Hille and Walsh [1] provides a thorough bibliography of the literature up to 1939.

The general theory of orthogonal polynomials begins with the work of Tchebichef [3] and Stieltjes [2]. The moment functional was introduced by M. Riesz [1] in connection with the problem of moments.

Favard's theorem was announced in Favard [1]. As to the other claims, see Shohat [2], Natanson [1, p. 349]. The extension of Favard's theorem to the quasi-definite case with the functional expressed in terms of a Stieltjes integral with an integrator of bounded variation appeared in Shohat [3]. See also Geronimus [3], [4].

This multiple discovery is not surprising since the theorem is really implicitly contained in the theory of continued fractions. It seems quite likely that mathematicians who worked with continued fractions were aware of the theorem (e.g. see Sherman [1]) but never bothered to formulate it explicitly. Nevertheless, the explicit formulation was a real contribution since most workers in orthogonal polynomials tend to avoid continued fractions whenever possible.

An extensive discussion of the recurrence formula in the general context of discrete boundary value problems is given by Atkinson [1]. Weak orthogonality is discussed by Struble [1] and by Krall and Sheffer [1] (to whom the name is due). An early study of specific examples of weak orthogonality is Szegö [4].

Most of the general results concerning the zeros given in this chapter go back to Stieltjes. More detailed theorems about the zeros, especially for the classical orthogonal polynomials, can be found in Szegö's book [5]. See also chapter IV of this book.

The Gauss quadrature formula is frequently called the Gauss-Jacobi quadrature formula. The original work of Gauss and Jacobi dealt with the case of an integral with a constant weight function (zeros of the Legendre polynomials as nodes). The formula is discussed in most books on approximation theory. In particular, see Freud [1], Krylov [1], Natanson [1], and Stroud and Secrest [2]. As for the Lagrange interpolation formula relative to orthogonal polynomials, see Freud [1], Natanson [1], and Szegö [5]. Quadrature formulas with equal weights ($A_{ni} = A_{nj}$; cf. Ex. 6.3) are called Tchebichef quadratures. Weight functions for which Tchebichef quadrature formulas are known to exist are quite rare—see Ullman [1].

The relations connecting the recurrence formulas for an OPS and its KPS appear in Stieffel [1]. The same relations appear in well known formulas of Stieltjes [2] which relate the coefficients in S-fractions and their associated J-fractions. They are implicit in the recurrence formulas of Karlin and McGregor (see Ex. 9.9). In the form given in §9 in which the recurrence formula for the corresponding symmetric OPS is also involved, they were given in Chihara [3]. Another study of kernel polynomials is Rossum [1].

We have not discussed the concept of "finite orthogonality" in which the moment functional $\mathcal{L}$ is positive-definite on a set of finite cardinality $N + 1$ but is not quasi-definite. In this case (an instance of weak orthogonality), a finite set of polynomials $p_i(x)$, $0 \leqslant i \leqslant N$, is uniquely determined such that $\mathcal{L}[p_i(x)p_j(x)] = \delta_{ij}$ ($i,j = 0, 1, \ldots, N$). Examples are the Hahn, Krawtchouk, and discrete Tchebichef polynomials mentioned in Chapter V. For further examples and applications see Atkinson [1], Cheney [1], Erdélyi [1], Forsythe [2], Hildebrand [1], Peck [1], Szegö [5], Wilson [4], as well as the references given in Chapter V.

Most of the known examples of such polynomials are related to well known positive-definite OPS by simple limiting relations. Neuts and Lambert [1] and Wilson [2] discuss general theorems dealing with such limits.

Exercises 4.15 and 7.11 illustrate a type of problem that is of current interest. See also Askey [3], [6], [8], Gasper [1], and Wilson [3].

The original *Turán inequality* for Legendre polynomials (see Szegö [6]) has been the inspiration for a great deal of interest in inequalities involving determinants whose elements are orthogonal polynomials. A somewhat trivial example is (4.13). See Askey [7], Danese [1], Forsythe [1], Gasper [2], Koschmieder [1], Mukherjee and Nanjundiah [1], Skovgaard [1], Szasz [1], Szegö [8]. The Turán type inequalities have been generalized to determinants like $D(n, r)$ in Exercise 7.12 and studied extensively by Karlin and Szegö [1]. Even more general determinants have been investigated by

Karlin [1], Karlin and McGregor [4], [6] and placed in the larger context of the theory of total positivity by Karlin [3].

Some other examples of positivity questions involving orthogonal polynomials are Askey and Gasper [1], Gasper [3], Karlin and McGregor [1, Theorem 3]. Askey [9] discusses a number of positivity problems and provides an extensive bibliography.

Expansion theory is not discussed in this book. For this most important aspect of the general theory, see Alexits [1], Freud [1], Kasczmarz and Steinhaus [1], Tricomi [3], Szegö [5].

For other approaches to the general theory of orthogonal polynomials, see Atkinson [1], Karlin and Shapley [1], Stone [1], Wall [1], Wynn [1]. For the important case of orthogonal polynomials in a complex variable, see Freud [1], Geronimus [5], [6], Grenander and Szegö [1], and Szegö [5]. For generalizations to "lacunary orthogonal polynomials" (polynomials in x^k), see Endl [1], [2], [3], [4], Frank [1], Rossum [2], Sallay [1].

CHAPTER II. Most of the ideas in this chapter can be traced back to Tchebichef [3] and Stieltjes [2]. Helly's theorems appeared in 1912. Stieltjes and Hamburger [1] worked with continued fractions and integrals of the form III-(4.8).

The discussion in §4 is rather incomplete without consideration of determinacy questions. Stieltjes [2] showed that if the Stieltjes moment problem is indeterminate, then $\xi_i < \xi_{i+1}$ and $\sum \xi_i^{-1} < \infty$. This forms the basis for the remark that $\mathcal{L}$ is determined if σ or τ is finite. This also shows that when the Stieltjes moment problem is indeterminate, the zeros of the orthogonal polynomials are rather simply distributed and the largest zeros grow rapidly. No precise result of this type seems to be known for the general case of a Hamburger moment problem. An indication that something similar must be true is given by Greenstein [1, p. 628].

A great deal of information regarding the spectrum is given by Stone [1]. In particular, in the determinate case, Theorem 4.4 (i) is due to Stone [1, Theorem 10.42]. Here also can be found a proof of the fact alluded to at the end of §5 that in the indeterminate case, every closed set is the derived set of the spectrum of some solution of the moment problem.

The Tchebichef inequalities were stated without proof by Tchebichef [3] in 1874 and first proved by Markov [1] in 1884. An independent proof was given almost simultaneously by Stieltjes. For references and historical remarks, see Krein [1].

Of special significance for the theory of orthogonal polynomials is the concept of *extremal solution* to an indeterminate Hamburger moment problem which is due to P. Nevanlinna. A theorem of M. Riesz [2] states that $\{P_n(x)\}$ is complete in the L_2 space determined by the measure $d\psi$ if and only if ψ is the solution of a determined Hamburger moment problem or else is an extremal solution of an indeterminate moment problem. For a discussion of these matters and references, see Ahiezer [1] or Shohat and Tamarkin [1]. See also Ahiezer and Krein [1].

The most general criterion for determinacy of a Hamburger moment problem is probably that due to Carleman (see Shohat and Tamarkin [1, p. 19]). A special case of Carleman's criterion is stated in Exercise 6.3. Hausdorff's solution of his moment problem appeared in 1923. Boas' theorem was announced in 1938 and extended by Polya who refers to earlier results of Borel. The proof given here is Boas'. For references, see Shohat and Tamarkin [1].

CHAPTER III. We refer to Wall [2] for references to original works on continued fractions.

For a study of J-fractions from the viewpoint of operators on a Hilbert space, see Stone [1]. The most thorough study of numerator polynomials is probably Sherman [1]. Later studies of numerator polynomials have largely concentrated on the properties of specific examples. However, see Dickinson [2], Goldberg [1], Maki [1], [2], Palama [1]. For a converse to the property (4.4), see W. A. Al-Salam [2]. For the related OPS described in Exercises 4.2 and 4.3, see Chihara [2] and Allaway [1].

Markov's theorem on the uniform convergence of (4.7) follows from an application of the Stieltjes-Vitali theorem. The Stieltjes inversion formula first appeared in Stieltjes' classic memoir [2]. For proofs, see Ahiezer [1], Wall [1].

Continued fractions—hence orthogonal polynomials—have important connections with the so-called Padé table. See the survey by Gragg [1] and the references given there.

The systematic development of the theory of chain sequences is due to Wall although sequences of this type appeared earlier in continued fraction theory. Most of the results in §4 and §5 can be found in Wall [1]. Most of the remainder appeared in Chihara [3], [5], [7]. For some additional results, see Haddad [1], [3].

Chain sequences are closely related to the *g-algorithm* in numerical analysis—see Bauer [1]. For connections with the Hausdorff moment problem, see Seall and Wetzel [1], [2]. Chain sequences have been generalized by Seall and Wetzel [2] and by Haddad [2].

CHAPTER IV. The study of orthogonal polynomials whose zeros all lie in $(0, \infty)$ begins properly with Stieltjes, of course. The explicit study of this case from the viewpoint of the recurrence formula was first made by Karlin and McGregor [1]. They write the recurrence formula in a special form (see I-Ex. 9.9) which has probabilistic interpretations and develop an extensive theory with significant applications. The application of chain sequences to the study of the recurrence formula was initiated in Chihara [3].

The special case $\sigma = 0$ of Theorem 3.5 appeared in Stieltjes [2] (pp. 560-566). Goldberg [1] has made a detailed study of this case. See also Dickinson, Pollack and Wannier [1]. Krein's generalization of Stieltjes' theorem can be found in Ahiezer and Krein [1] (pp. 230-231). Using continued fraction theory, Maki [1] proved that if $\lambda_n \to 0$ and $\{c_n\}$ has finitely many sequential limit points, then the derived set of the spectrum consists of these limit points. Maki [3] subsequently proved that if $\lambda_n \to 0$, then every limit point of $\{c_n\}$ is a spectral point and conjectured that c is a limit point of $\{c_n\}$ if and only if c is a limit point of the spectrum. This conjecture has been validated (Chihara [9]). See also Allaway [2] for additional related results.

Theorem 3.6 and the special case (3.7) first appeared in Chihara [5] and [3], respectively, but with the added hypothesis that the Hamburger moment problem was determined. Maki [2] proved a theorem about continued fractions that showed (3.7) is valid without assumptions about the moment problem.

Blumenthal's theorem appeared in his Göttingen thesis [1]. The generalization and the extension of Blumenthal's theorem appeared in Chihara [7]. For an interesting (and somewhat surprising) result concerning an analogue of Blumenthal's theorem for polynomials orthogonal on a Jordon curve in the complex plane, see Duren [1]. Not discussed in this book are important related theorems which predict the density and distribution of zeros on the basis of conditions on the weight function. See Szegö [5, Chapter 6], Freud [2], [3], and Ullman [2].

To a certain extent, determinacy questions for the associated moment problems can also be decided from the behavior of the coefficients in the recurrence formula (Chihara [5], [8]).

It has probably not escaped the reader's attention that there is practically no information in this book concerning the case $\sigma = -\infty$, $\tau = \infty$. Of course, in the symmetric case, the relations in I-Theorem 9.1 can be used and results about the case $\sigma = 0$, $\tau = \infty$ transformed to the symmetric case. Aside from this more or less trivial case, little appears known and many questions suggest themselves.

In particular, what happens if $c_n \to \infty$ and $\lambda_{n+1}/(c_n c_{n+1}) \to L > 1/4$ beyond the fact that $(\sigma, \tau) = (-\infty, \infty)$? We offer the conjecture that in general, if $c_n \to \infty$ and $L \geqq 1/4$, then the zeros are dense in (σ, ∞). The lack of examples here is strongly felt. In fact, the only examples we know of in which $c_n \to \infty$ and $L > 1/4$ are the Meixner polynomials of the second kind and their generalizations by Pollaczek (see VI-§3 and §4). In these examples, the zeros *are* dense in $(-\infty, \infty)$.

APPENDIX

Table of Recurrence Formulas

General Remarks

Since the three term recurrence formulas satisfied by an OPS $\{P_n(x)\}$ and the essentially equivalent OPS $\{K_n P_n(ax + b)\}$ can look quite different, the most effective way to check the identity of an OPS given by its recurrence formula is to consider the corresponding monic polynomials.

If $\{Q_n(x)\}$ satisfies

$$A_n Q_{n+1}(x) = (B_n x - C_n) Q_n(x) - D_n Q_{n-1}(x),$$

then the monic polynomials

$$\hat{Q}_n(x) = \frac{A_0 A_1 \cdots A_{n-1}}{B_0 B_1 \cdots B_{n-1}} Q_n(x)$$

satisfy

$$\hat{Q}_{n+1}(x) = (x - \gamma_n)\hat{Q}_n(x) - \delta_n \hat{Q}_{n-1}(x)$$

where

$$\gamma_n = \frac{C_n}{B_n}, \qquad \delta_n = \frac{A_{n-1} D_n}{B_{n-1} B_n}.$$

Thus

$$Q_n(x) = \frac{a^{-n} B_0 \cdots B_{n-1}}{A_0 A_1 \cdots A_{n-1}} \hat{P}_n(ax + b),$$

where

$$\hat{P}_{n+1}(x) = (x - \alpha_n)\hat{P}_n(x) - \beta_n \hat{P}_{n-1}(x), \tag{1}$$

and

215

$$\frac{C_n}{B_n} = \frac{\alpha_n - b}{a}, \qquad \frac{A_{n-1} D_n}{B_{n-1} B_n} = \frac{\beta_n}{a^2}.$$

The following table lists the coefficients α_n and β_n corresponding to (1) for all positive-definite OPS mentioned in this book for which the recurrence formulas are known. The entries listed under α_n and β_n are assumed valid for $n \geq 0$ and $n \geq 1$, respectively, except where noted otherwise. For the sake of brevity, positivity conditions on any parameters that appear are omitted.

It may be noted from (I) of the table that when $\alpha_n = an + b$ and $\beta_n = cn^2 + dn + f > 0$, the corresponding positive-definite OPS can be identified and the orthogonality relations explicitly found in all cases except: (i) $a = c = 0$, $df \neq 0$, (ii) $c = 0$, $adf \neq 0$, (iii) $a^2 \geq 4c$, $f \neq 0$. (The case $a^2 < 4c$, $f \neq 0$ can be reduced to the form of the polynomials of Pollaczek of VI-(5.13) but there are certain exceptional values of the parameters for which $\beta_n > 0$ but Pollaczek's conditions for orthogonality are not met.)

Table of Recurrence Formulas

	α_n	β_n	$\hat{P}_n(x)$
I	α_n constant or linear in n; β_n constant, linear or quadratic in n.		
1	0	1/4	$\hat{U}_n(x)$ (Tchebichef, 2nd kind) I-Ex. 4.9
2	0 except:	1/4 except:	$\hat{T}_n(x)$ (Tchebichef, 1st kind) I-(4.5)
(a)		$\beta_1 = 1/2$	
(b)		β_1 arbitrary	VI-(13.10) (i)
(c)	$\alpha_0 \neq 0$		VI-(13.10) (iii)
(d)	$\alpha_0 \neq 0$	$\beta_1 = 1/2$, $\beta_2 = \dfrac{1 - \beta_1}{2}$	VI-(13.10) (ii)
(e)			VI-(13.10) (iv)
3	0	$n/2$	$\hat{H}_n(x)$ (Hermite) V-(2.28)
4	$-n$	$\beta > 0$	VI-(6.11)
5	$n + a$	an	$C_n^{(a)}(x)$ (Charlier) VI-(1.4)
6	$2n + \alpha + 1$	$n(n + \alpha)$	$\hat{L}_n^\alpha(x)$ (Laguerre) V-(2.27)

217

	α_n	β_n	
7	$\dfrac{(1+c)n + \beta c}{1-c}$	$\dfrac{c}{(1-c)^2}n(n+\beta-1)$	$\hat{m}_n(x; c, \beta)$ (Meixner, 1st kind) VI-(3.2)
8	$(2n+\eta)\delta$	$(\delta^2+1)n(n+\eta-1)$	$M_n(x; \delta, \eta)$ (Meixner, 2nd kind) VI-(3.17)
9	$-(\cot\phi)(n+\lambda+c)$	$1/4(\csc^2\phi)(n+c)(n+2\lambda+c-1)$	VI-(5.13)
II	α_n quadratic in n and β_n quadratic in n		
1	$(2n+1)^2 + (2n)^2 k^2$	$4n^2(2n-1)^2 k^2$	VI-(9.10)
2	$(2n)^2 + (2n+1)^2 k^2$	$4n^2(2n-1)^2 k^2$	VI-(9.12)
3	$(2n+2)^2 + (2n+1)^2 k^2$	$4n^2(2n+1)^2 k^2$	VI-(9.13)
4	$(2n+1)^2 + (2n+2)^2 k^2$	$4n^2(2n+1)^2 k^2$	VI-(9.11)
5	$(2n+1)^2 + (2n+1)^2 k^2$	$4n^2(4n^2-1)k^2$	VI-(9.1)
6	$(2n+2)^2 + (2n+2)^2 k^2$	$4n(n+1)(2n+1)^2 k^2$	VI-(9.2)
7	$2(2n)^2 + (4n+1)\eta$	$2n(2n-1)(2n+\eta-2)(2n+\eta-1)$	VI-(3.24)
8	$2(2n+1)^2 + (4n+3)\eta$	$2n(2n+1)(2n+\eta-1)(2n+\eta)$	VI-(3.24)
III	α_n and β_n both rational functions of n		
1	0	$\dfrac{1}{4(n+\nu)(n+\nu-1)}$	$\hat{h}_{n,\nu}(x)$ (modified Lommel) VI-(6.2)

	0 except:		
2	Same as 1.	$\alpha_0 \neq 0$	VI-(6.9)
3	$\dfrac{n}{(n+\alpha)(n+\alpha-1)}$	0	VI-(7.6)
4	$\dfrac{n^2}{(2n-1)(2n+1)}$	0	$\hat{P}_n(x)$ (Legendre) I-Ex. 1.6
5	$\dfrac{(n+\nu)^2}{(2n+2\nu-1)(2n+2\nu+1)}$	0	$\hat{P}_n(x,\nu)$ (Associated Legendre) VI-(12.1)
6	$\dfrac{n(n+2\lambda-1)}{4(n+\lambda)(n+\lambda-1)}$	0	$\hat{P}_n^\lambda(x)$ (Gegenbauer) V-(2.5)
7	$\dfrac{n^2(n^2-\lambda^2)}{(2n-1)(2n+1)}$	0	VI-(8.7)
8	$\dfrac{(n+c)(n+2\lambda+c-1)}{4(n+\lambda+a+c-1)(n+\lambda+a+c)}$	$\dfrac{-b}{n+\lambda+a+c}$	$\hat{P}_n^\lambda(x;a,b,c)$ (Pollaczek) VI-(5.8)
9	$\dfrac{(\alpha n+\alpha+1)(\alpha n-\alpha+1)}{4(\alpha n+1)^2}$	$\dfrac{\alpha^2}{2(\alpha n+1)(\alpha n+\alpha+1)}$, $\quad \alpha_0 = -\dfrac{2\alpha+1}{2\alpha+2}$	IV-Ex. 2.14
10	$\dfrac{n(n+c)}{4(n+a)^2}$	$-\dfrac{(n+a)b-\gamma a}{2(n+a)(n+a+1)}$, $\quad a\gamma^2+b\gamma+a-c=0$	VI-(5.18)

11 $\dfrac{\beta^2 - \alpha^2}{(2n+\alpha+\beta)(2n+\alpha+\beta+2)}$ $\dfrac{4n(n+\alpha)(n+\beta)(n+\alpha+\beta)}{(2n+\alpha+\beta-1)(2n+\alpha+\beta)^2(2n+\alpha+\beta+1)}$ $P_n^{(\alpha,\beta)}(x)$ (Jacobi) V-(2.26)

(if $\alpha = -\beta, \alpha_0 = -\alpha$)

IV	α_n and β_n exponential functions of n		
1	$(1+a)q^n$	$-a(1-q^n)q^{n-1}$	VI-(10.5)
2	$(1+a)q^{-n}$	$aq^{1-2n}(1-q^n)$	VI-(10.6)
3	$[b+q-b(1+q)q^n]q^n$	$b(1-q^n)(1-bq^{n-1})q^{2n}$	$W_n(x;q,b)$ (Wall) VI-(11.1)
4	$-[p+q-(1+q)q^{-n}]q^{-n-3/2}$	$(1-q^n)(1-pq^{n-1})q^{-4n}$	$S_n(x;q,p)$ (Stieltjes-Wigert) VI-(2.9)

V	Miscellaneous		
1	0	$\beta_{2m-1}=a,\ \beta_{2m}=b$	III-Ex. 4.4
2	0	$\beta_{2m-1}=a,\ \beta_{2m}=m$	VI-(1.13)
3	0	$\beta_{2m-1}=m+\mu-\tfrac{1}{2},\ \beta_{2m}=m$	$\hat{H}_n^\mu(x)$ (generalized Hermite) V-(2.46)
4	0	$\beta_{2m-1}=\dfrac{c}{1-c}(m+\beta-1)$, $\ \beta_{2m}=\dfrac{m}{1-c}$	VI-(3.16)
5	0	$\beta_{2m-1}=(2m-1)^2$, $\ \beta_{2m}=(2m)^2k^2$	VI-(9.3)

6	0	$\beta_{2m-1} = (2m-1)^2 k^2$ $\beta_{2m} = (2m)^2$	VI-(9.4)
7	0	$\beta_{2m-1} = \dfrac{(m+\beta)(m+\alpha+\beta)}{(2m+\alpha+\beta-1)(2m+\alpha+\beta)}$ $\beta_{2m} = \dfrac{m(m+\alpha)}{(2m+\alpha+\beta)(2m+\alpha+\beta+1)}$	V-(2.41)
8	$(-1)^n c$	β_n as in 7	V-(2.42)
9	0	$\beta_{2m-1} = (1 - pq^{m-1})q^{-2m+1/2}$ $\beta_{2m} = (1 - q^m)q^{-2m-1/2}$	VI-(2.14)
10	0	$\beta_{2m-1} = (1 - bq^{m-1})q^m$ $\beta_{2m} = b(1 - q^m)q^m$	VI-(11.11)
11	$\alpha_{2m-1} = -(b-q)q^{m-1}$ $\alpha_{2m} = -(1-b)q^m$	$\beta_{2m-1} = b(1 - bq^{m-1})q^{m-1}$ $\beta_{2m} = (1 - q^m)q^m$	VI-(11.15)
12	0	$\beta_{3m-1} = 1,\ \beta_{3m} = \dfrac{m}{2m+1}$ $\beta_{3m+1} = \dfrac{m+1}{2m+1}$	VI-(13.20)
13	0	$\dfrac{n^2}{(2n-1)(2n+1)(1 - cf_{n-1})(1 - cf_n)}$	VI-(13.22)

List of Frequently Used Symbols

$b_n{}^{(k)}$	b_{n+k}, p. 94
$\det(a_{ij})$	determinant of (a_{ij})
$F(a,b;c;x)$	hypergeometric function
${}_pF_q(a_i;c_j;x)$	generalized hypergeometric function
$J_\nu(x)$	Bessel function
$\mathscr{L}$	moment functional, p. 6
$\mathscr{L}_\kappa^*$	moment functional for kernel polynomials, p. 35
m_k	kth minimal parameter, p. 93
M_k	kth maximal parameter, p. 94
$\hat{P}_n(x)$	monic form of $P_n(x)$, p.10
$P_n^{(k)}(x)$	numerator polynomial of order k, p. 86
$P_n^*(\kappa,x)$	kernel polynomial, p. 35
sgn x	signum function, p. 28
x_{nk}	kth zero of $P_n(x)$, p. 28
X	set of zeros of $\left\{P_n(x)\right\}$, p. 61
$\Gamma(x)$	gamma function
δ_{mn}	Kronecker delta, p. 5
Δ	difference operator, p. 160
Δ_n	$\det(\mu_{i+j})$, p. 11
η_j	$\lim_{n\to\infty} x_{n,n-j+1}$, p. 29
μ_n	nth moment of $\mathscr{L}$, p. 7
ξ_i	$\lim_{n\to\infty} x_{ni}$, p. 29
σ	$\lim_{i\to\infty} \xi_i$, p. 61
τ	$\lim_{j\to\infty} \eta_j$, p. 61
ψ	distribution function, p. 51

223

$\mathfrak{S}(\psi)$ spectrum of ψ, p. 51

$[x]$ greatest integer $\leq x$

$(a)_n$ p. 150

$[a]_n$ p. 173

$\binom{a}{n}$ binomial coefficient

Bibliography

Abdul-Halim, N. and Al-Salam, W. A.

1 *A characterization of the Laguerre polynomials*, Rendiconti del Seminario Mat. Univ. Padova, 34 (1964), 176-179.

Ahiezer, N. I. (Achyeser, Akhiezer)

1 *Über eine Eigenschaft der "elliptischen" Polynome*, Commun. de la Soc. Math. Kharkoff, 4 (1934), 3-8.
2 *Orthogonal polynomials on several intervals*, Soviet Math. Doklady, 1 (1960), 989-992.
3 *The Classical Moment Problem*, Hafner Publ. Co., N.Y., 1965.

Ahiezer, N. I. and Krein, M. G.

1 *Some Questions in the Theory of Moments*, Translations of Mathematical Monographs, Vol. 2, Amer. Math. Soc., Providence, 1962.

Alexits, G.

1 *Convergence Problems of Orthogonal Series*, Pergamon Press, Oxford, 1961.

Allaway, W. R.

1 *The identification of a class of orthogonal polynomial sets*, Thesis, U. of Alberta, 1972.
2 *On finding the distribution function for an orthogonal polynomial set*, Pacific J. Math., 49 (1973), 305-310.

Al-Salam, N. A.

1 *Orthogonal polynomials of hypergeometric type*, Duke Math. J., 33 (1966), 109-122.

Al-Salam, W. A.

1 *The Bessel polynomials*, Duke Math. J., 24 (1957), 529-546.
2 *On a characterization of orthogonality*, Math. Magazine, 3 (1957/1958), 41-44.
3 *Some functions related to the Bessel polynomials*, Duke Math. J., 26 (1959), 519-540.
4 *On the orthogonality of some systems of polynomials*, Memoria Publicada en Collectanea Mathematica, XVI (1964), 187-192.
5 *Characterization of certain classes of orthogonal polynomials related to elliptic functions*, Annali di Matematica pura et applicata, LXVII (1965), 75-94.

Al-Salam, W. A. and Carlitz, L.

1 *Bernoulli numbers and Bessel polynomials*, Duke Math. J., 26 (1959), 437-446.
2 *Some orthogonal q-polynomials*, Math. Nachr., 30 (1965), 47-61.

Al-Salam, W. A. and Chihara, T. S.

1 *Another characterization of the classical orthogonal polynomials*, SIAM J. Math. Anal., 3 (1972), 65-70.

Angelescu, A.

1 *Sur les polynomes orthogonaux en rapport avec d'autres polynomes*, Bull. de la Soc. de Sc. de Cluj Roumanie, 1 (1921-23), 44-59.

Askey, R. A.

1 *Orthogonal expansions with positive coefficients*, Proc. Amer. Math. Soc., 16 (1965), 1191-1194.
2 *Norm inequalities for some orthogonal series*, Bull. Amer. Math. Soc., 72 (1966), 808-823.
3 *Jacobi polynomial expansions with positive coefficients and imbeddings of projective spaces*, Bull. Amer. Math. Soc., 74 (1968), 301-304.
4 *A transplantation theorem for Jacobi series*, Illinois J. Math., 13 (1969), 583-590.
5 *Mean convergence of orthogonal series and Lagrange interpolation*, Stichting Math. Centrum, Amsterdam, Afd. Toegepaste Wisk., TW 113, (1969), 23 pp.
6 *Orthogonal polynomials and positivity*, in *Special Functions and Wave Propagation*, D. Ludwig and F. W. J. Olver (Ed.), Studies in Applied Mathematics, 6, SIAM, Philadelphia, (1970), 64-85.
7 *An inequality for the classical polynomials*, Koninkl. Nederl. Akad. Wetenschappen - Amsterdam, Proc., A, 73 (1970), 22-25.
8 *Orthogonal expansions with positive coefficients, II*, SIAM J. Math. Anal., 2 (1971), 340-346.
9 *Orthogonal Polynomials and Special Functions*, Society for Industrial and Applied Mathematics, Philadelphia, 1975.

Askey, R. A. and Gasper G.

1 *Certain rational functions whose power series have positive coefficients*, Amer. Math. Monthly, 79 (1972), 327-341.

Askey, R. A. and Hirschman, I. I.

1 *Mean summability for ultraspherical polynomials*, Math. Scandia 12 (1963), 167-177.

Askey, R. A. and Wainger, S.

1 *On the behavior of special classes of ultraspherical expansions* I, II., Journal d'Analyse Math. XV (1965), 193-220; 221-244.
2 *Mean convergence of expansions in Laguerre and Hermite series*, Amer. J. Math., 87 (1965), 695-708.

Atkinson, F. V.

1 *Discrete and Continuous Boundary Problems*, Academic Press, N.Y., 1964.

Barrucand, P. and Dickinson, D.

1 *On cubic transformations of orthogonal polynomials*, Proc. Amer. Math. Soc., 17 (1966), 810-814.
2 *On the associated Legendre polynomials*, in *Orthogonal Expansions and Their Continual Analogues*, D. T. Haimo (Ed.), Southern Illinois U. Press, Edwardsville, 1968, pp. 43-50.

Bateman, H.

1 *An orthogonal property of the hypergeometric polynomial*, Proc. Nat. Acad. Sci., U.S.A., 28 (1942), 374-377.

Bauer, F. L.

1 *The g-algorithm*, J. Soc. Indust. Appl. Math., 18 (1960), 1-17.

Beale, F. S.

1 *On a certain class of orthogonal polynomials*, Ann. Math. Statist., 12 (1941), 97-103.

Bernstein, S.

1 *Sur une classe de polynomes orthogonaux*, Comm. de la Soc. Math. Kharkoff, 4 (1930), 79-93; *Complémant*, Ibid., 5 (1932), 59-60.

Blumenthal, O.

1 *Über die Entwicklung einer willkürlichen Funktion nach den Nennern des Kettenbruches für* $\int_{-\infty}^{0} [\phi(\xi)/(z-\xi)]d\xi$. Dissertation, Göttingen, 1898.

Boas, R. P. and Buck, R. C.

1 *Polynomial Expansions of Analytic Functions*, Springer-Verlag, Berlin, 1964.

Bochner, S.

1 *Über Sturm-Liouvillesche Polynomsysteme*, Math. Zeit. 29 (1929), 730-736.

Boyer, R. H.

1 *Discrete Bessel functions*, J. Math. Anal. Appl., 2 (1961), 509-624.

Brafman, F.

1 *On Touchard polynomials*, Canad. J. Math., 9 (1957), 191-193.

Brenke, W. C.

1 *On generating functions of polynomial systems*, Amer. Math. Monthly, 52 (1945), 297-301.

Burchnall, J. L.

1 *The Bessel polynomials*, Canad. J. Math., 3 (1951), 62-68.

Burchnall, J. L. and Chaundy, T. W.

1 *Commutative ordinary differential operators II. The identity $P^m = Q^m$*, Proc. Roy. Soc., Ser. A., 134 (1931), 471-485.

Carlitz, L.

1 *Polynomials related to theta functions*, Ann. Mat. Pura Appl., (4) 4 (1955), 359-373.
2 *Some polynomials of Touchard connected with the Bernoulli numbers*, Canad. J. Math., 9 (1957), 188-190.
3 *On some polynomials of Tricomi*, Boll. Un. Mat. Ital. (3) 13 (1958), 58-64.
4 *Note on orthogonal polynomials related to theta functions*, Publ. Math. Debrecen, 5 (1958), 222-228.
5 *Bernoulli and Euler numbers and orthogonal polynomials*, Duke Math. J., 26 (1959), 1-16.
6 *Some orthogonal polynomials related to elliptic functions*, Ibid., 27 (1960), 443-460.
7 *Some orthogonal polynomials related to elliptic functions, II. Arithmetic properties*, Ibid., 28 (1961), 107-124.
8 *Some orthogonal polynomials related to the Fibonacci numbers*, Fibonacci Quart., 4 (1966), 43-48.
9 *The generating function for the Jacobi polynomial*, Rendiconti Seminario Mat. Univ. Padova, 38 (1967), 86-88.

Charlier, C. V. L.

1 *Über die darstellung willkürlicher Functionen*, Arkiv för matematik, astronomi och fysik, 2 (1905-1906), No. 20, 35 pp.

Cheney, E. W.

1 *Introduction to Approximation Theory*, McGraw-Hill Book Co., N.Y., 1966.

Chihara, T. S.

1 *Generalized Hermite polynomials*, Thesis, Purdue, 1955.
2 *On co-recursive orthogonal polynomials*, Proc. Amer. Math. Soc., 8 (1957), 899-905.
3 *Chain sequences and orthogonal polynomials*, Trans. Amer. Math. Soc., 104 (1962), 1-16.
4 *On kernel polynomials and related systems*, Boll. Un. Mat. Ital., (3) 19 (1964), 451-459.
5 *On recursively defined orthogonal polynomials*, Proc. Amer. Math. Soc., 16 (1965), 702-710.
6 *Orthogonal polynomials with Brenke type generating functions*, Duke Math. J., 35 (1968), 505-518.
7 *Orthogonal polynomials whose zeros are dense in intervals*, J. Math. Anal. Appl., 24 (1968), 362-371.
8 *On indeterminate Hamburger moment problems*, Pacific J. Math., 27 (1968), 475-484.
9 *The derived set of the spectrum of a distribution function*, Pacific J. Math., 35 (1970), 571-574.
10 *A characterization and a class of distribution functions for the Stieltjes-Wigert polynomials*, Canad. Math. Bull, 13 (1970), 529-532.
11 *The orthogonality of a class of Brenke polynomials*, Duke Math. J. 38 (1971), 599-603.

Churchill, R. V.

1 *Fourier Series and Boundary Value Problems*, McGraw-Hill Book Co., N.Y., 2nd Ed., 1960.

Coulton, D.

   1 *Jacobi polynomials of negative index and a nonexistence theorem for the generalized axially symmetric potential equation*, SIAM J. Appl. Math., 16 (1968), 771-776.

Courant, R. and Hilbert, D.

   1 *Methods of Mathematical Physics*, Vol. 1, Interscience Publ., N.Y., 3rd Ed., 1953.

Cramér, H.

   1 *Mathematical Methods of Statistics*, Princeton U. Press, Princeton, 1946.

Cryer, C. W.

   1 *Rodrigues' formula and the classical orthogonal polynomials*, Boll. Un. Mat. Ital., (3) 25 (1970), 1-11.

Danese, A. E.

   1 *Some identities and inequalities involving ultraspherical polynomials*, Duke Math. J., 26 (1959), 349-360.

Davis, P. J.

   1 *Interpolation and Approximation*, Blaisdell Publ. Co., Waltham, 1963.

Dickinson, D. J.

   1 *On Lommel and Bessel polynomials*, Proc. Amer. Math., 5 (1954), 946-956.
   2 *On certain polynomials associated with orthogonal polynomials*, Boll. Un. Mat. Ital., (3) 13 (1958), 116-124.
   3 *On quasi-orthogonal polynomials*, Proc. Amer. Math. Soc., 12 (1961), 185-194.

Dickinson, D. J., Pollack, H. O. and Wannier, G. H.

   1 *On a class of polynomials orthogonal over a denumerable set*, Pacific J. Math., 6 (1956), 239-247.

Dickinson, D. and Warsi, S. A.

   1 *On a generalized Hermite polynomial and a problem of Carlitz*, Boll. Un. Mat. Ital., 18 (1963), 256-259.

Duffin, R. J. and Schmidt, T. W.

   1 *An extrapolator and scrutator*, J. Math. Anal. Appl., 1 (1960), 215-227.

Duren, P. L.

   1 *Polynomials orthogonal over a curve*, Michigan Math. J., 12 (1965), 313-316.

Eagleson, G. K.

   1 *A duality relation for discrete orthogonal systems*, Studia Scientiarum Mathematicarum

Hungaria, 3 (1968), 127-136.

2 *A characterization theorem for positive definite sequences on the Krawtchouk polynomials,* Austral. J. Statist., 2 (1969), 29-38.

Ebert, R.

1 *Über Polynomsysteme mit Rodriguesscher Darstellung,* Dissertation, Cologne, 1964.

Endl, K.

1 *Sur une classe de polynomes orthogonaux généralisant ceux de Laguerre et de Hermite,* C.R. Acad. Sci. Paris, 241 (1955), 723-724.

2 *Orthogonalisierung auf einem k-strahliegen symmetrischen integrationsstern,* Math. Zeit., 65 (1956), 1-6.

3 *Über eine ausgezeichnete Eigenschaft der Koeffizientenmatrizen des Laguerreschen und des Hermiteschen Polynomsystems, Ibid.,* 65 (1956), 7-15.

4 *Les polynomes de Laguerre et de Hermite comme cas particuliers d'une classe de polynomes orthogonaux,* Ann. Sci. École Norm. Sup., (3) LXXIII, (1956), 1-13.

Erdélyi, A. (with Magnus, A., Oberhettinger, F. and Tricomi, F).

1 *Higher Transcendental Functions,* McGraw-Hill Book Co., N.Y., Vol. 1, 2, 1953, vol. 3, 1955.

Favard, J.

1 *Sur les polynomes de Tchebicheff,* C.R. Acad. Sci. Paris, 200 (1935), 2052-2053.

Feldheim, E.

1 *On a system of orthogonal polynomials associated with a distribution of Stieltjes type,* C.R. (Doklady) Acad. Sci. URSS, 31 (1941), 528-533.

Feller, W.

1 *An Introduction to Probability Theory,* Vol. 1, John Wiley and Sons, N.Y., 1966.

Forsythe, G. E.

1 *Second order determinants of Legendre polynomials,* Duke Math. J., 18 (1951), 361-371.

2 *Generation and use of orthogonal polynomials for data fitting with a digital computer,* J. Soc. Industr. Appl. Math., 5 (1957), 74-88.

Frank, E.

1 *Orthogonality properties of C-fractions,* Bull. Amer. Math. Soc., 55 (1949), 384-390.

Freud, G.

1 *Orthogonal Polynomials,* Pergamon Press, Oxford, 1971.

2 *On the greatest zero of an orthogonal polynomial. I.* Acta Sci. Math. (Szeged), 34 (1973), 91-97.

3 *On the greatest zero of an orthogonal polynomial. II.* Acta Sci. Math. (Szeged), 36 (1974), 49-54.

Gasper, G.

1 *Linearization of the product of Jacobi polynomials, I, II*, Canad. J. Math., XXII (1970), 171-175; 582-593.
2 *An inequality of Turan type for Jacobi polynomials*, Proc. Amer. Math. Soc., 32 (1972), 435-439.
3 *Banach algebras for Jacobi series and positivity of a kernel*, Annals of Math., 95 (1972), 261-280.

Geronimus, J.

1 *On a set of polynomials*, Ann. Math., 31 (1930), 681-686.
2 *On a class of Appell polynomials*, Commun. de la Soc. Math. Kharkoff, (4) 8 (1934), 13-23.
3 *On polynomials orthogonal with regard to a given sequence of numbers*, Ibid., (4) 17 (1940), 3-18.
4 *The orthogonality of some systems of polynomials*, Duke Math. J., 14 (1947), 503-510.
5 *Polynomials orthogonal on a circle and their applications*, Amer. Math. Soc. Translations, No. 104 (1954), Reprinted as Amer. Math. Soc. Translations, Series 1, Vol. 3 (1962), 1-78.
6 *Polynomials Orthogonal on a Circle and Interval*, Pergamon Press, Oxford, 1960.

Goldberg, J. L.

1 *Polynomials orthogonal over a denumerable set*, Pacific J. Math., 15 (1965), 1171-1186.

Gottlieb, M. J.

1 *Concerning some polynomials orthogonal on a finite or enumerable set of points*, Amer. J. Math., 60 (1938), 453-458.

Gragg, W. B.

1 *The Padé table and its relation to certain algorithms of numerical analysis*, SIAM Review, 14 (1972), 1-62.

Greenleaf, H. E. H.

1 *Curve approximation by means of functions analogous to the Hermite polynomials*, Ann. Math. Statist., 3 (1932) 204-256.

Greenstein, D. S.

1 *On redundancy of powers and the moment problem*, Proc. Amer. Math. Soc., 13 (1962), 625-630.

Grenander, U. and Szegö, G.

1 *Toeplitz Forms and their Applications*, U. California Press, Berkeley, 1958.

Griñspun, Z. S.

1 *On a class of orthogonal polynomials*, [Russian] Vestnik Leningradskogo Universiteta Seria Matematiki, Mekhaniki i Astronomii, 21 (1966), 147-149.

232   AN INTRODUCTION TO ORTHOGONAL POLYNOMIALS

Haddad, H.

1 *Convergent Chain Sequences*, Thesis, U. Texas, 1962.
2 *Chain functions, A generalization of chain sequences*, Bull. College of Science, U. Baghdad, 9 (1966), 191-196.
3 *Chain sequence preserving linear transformations*, Ann. Scuola Norm. Sup. Pisa, III, 24 (1970), 78-84.

Hahn, W.

1 *Über die Jacobischen Polynome und zwei verwandte Polynomklassen*, Math. Zeit., 39 (1935), 634-638.
2 *Über höhere Ableitungen von Orthogonalpolynomen*, Ibid., 43 (1937), 101.
3 *Über Orthogonalpolynome mit drei Parametern*, Deutsche Math., 5 (1940), 273-278.
4 *Über Orthogonalpolynome die q-Differenzengleichungen genügen*, Math. Nachr., 2 (1949), 4-34.
5 *Über Polynome, die gleichzeitig zwei verschiedenen Orthogonalsystemen angehören*, Ibid., 2 (1949), 263-278.
6 *Beitrag zur Theorie der Heineschen Reihen*, Ibid., 2 (1949), 340-379.

Haimo, D. T. (Ed.)

1 *Orthogonal Expansions and their Continuous Analogues*, Southern Illinois U. Press, Edwardsville, 1968.

Hamburger, H.

1 *Über eine Erweiterung des Stieltjesschen Momentenproblems*, Math. Ann., 81 (1920), 235-319; 82 (1921), 120-164; 168-187.

Hammer, P. C. and Wicke, H. H.

1 *Quadrature formulas involving derivatives of the integrand*, Math. Tables and Other Aids to Comp., 14 (1960), 3-7.

Hardy, G. H.

1 *Notes on special systems of orthogonal functions (III): A system of orthogonal polynomials*, Proc. Cambridge Philos. Soc., 36 (1940), 1-8.

Heine, E.

1 *Handbuch der Kugelfunktionen*, 2 Vols, 2nd Ed., Berlin 1878, 1881.

Hermite, C.

1 *Sur un nouveau développment en série de fonctions*, C.R. Acad. Sci. Paris, 58 (1864), 93-100; 266-273.

Hewitt, E.

1 *Remarks on orthonormal sets in $L_2(a,b)$*, Amer. Math. Monthly, 61 (1954), 249-250.

Hildebrand, F. B.

1 *Introduction to Numerical Analysis*, McGraw-Hill Book Co., N.Y., 1956.

Hildebrandt, E. H.

1 *Systems of polynomials connected with the Charlier expansions and the Pearson differential and difference equation*, Ann. Math. Statist., 2 (1931), 379-439.

Hille, E.

1 *Analytic Function Theory*, Vol. 2, Blaisdell Publ. Co., Waltham, 1962.

Hobson, E. W.

1 *The Theory of Spherical and Ellipsoidal Harmonics*, Cambridge U. Press, Cambridge, 1931.

Humbert, P.

1 *Sur deux polynomes associés aux polynomes de Legendre*, Bull. Soc. Math. de France, XLVI (1918), 120-151.

Indritz, J.

1 *Methods in Analysis*, Macmillan Co., N.Y., 1963.

Jackson, D.

1 *Fourier Series and Orthogonal Polynomials*, (Carus Monograph No. 6), Math. Assoc. Amer., 1941.

Jacobi, C. G. J.

1 *Untersuchungen über die Differentialgleichung der hypergeometrischen Reihe*, J. Reine Angew. Math., 56 (1859), 149-165.

Karlin, S.

1 *Determinants of Eigenfunctions of Sturm-Liouville Equations*, J. d'Analyse Math., IX (1961/1962), 365-397.
2 *Sign regularity properties of classical orthogonal polynomials*, in *Orthogonal Polynomials and Their Continuous Analogues*, D. T. Haimo (Ed.), Southern Illinois U. Press, 1968, 55-74.
3 *Total Positivity*, Vol. I, Stanford U. Press, Stanford, 1962.

Karlin, S. and McGregor, J.

1 *The differential equations of birth and death processes and the Stieltjes moment problem*, Trans. Amer. Math. Soc., 85 (1957), 489-546.
2 *The classification of birth and death processes*, Ibid., 86 (1957), 366-401.
3 *Linear growth, birth and death processes*, J. Math. Mech., 7 (1958), 643-662.
4 *Coincidence properties of birth and death processes*, Pacific J. Math., 9 (1959), 1109-1140.
5 *The Hahn polynomials, formulas and an application*, Scripta Math., 26 (1961), 33-46.
6 *Determinants of orthogonal polynomials*, Bull. Amer. Math. Soc., 68 (1962), 204-209.
7 *On a genetics model of Moran*, Proc. Cambridge Philos. Soc., 58 (1962), 299-311.
8 *On some stochastic models in genetics*, in *Stochastic Models in Medicine and Biology*, J. Curland (Ed.), U. Wisconsin Press, Madison, 1964, 245-279.

Karlin, S. and Shapley, L. S.

1 *Geometry of Moment Spaces*, Amer. Math. Soc. Memoir No. 12, Amer. Math. Soc., Providence, 1953.

Karlin, S. and Studden, W. J.

1 *Tchebycheff Systems: with applications in analysis and statistics*, Interscience Publ., N.Y., 1966.

Karlin, S. and Szegö, G.

1 *On certain determinants whose elements are orthogonal polynomials*, J. d'Analyse Math., 8 (1961), 1-157.

Kasczmarz, S. and Steinhaus, H.

1 *Theorie der Orthogonalreihen*, Warsaw-Lwow, 1935, 2nd Ed. reprinted, Chelsea Publ. Co., N.Y., 1951.

Kellogg, O. D.

1 *Foundations of Potential Theory*, Berlin, 1929, Reprinted, Dover Publ., N.Y., 1953.

Konoplev, V. P.

1 *On orthogonal polynomials with weight functions vanishing or becoming infinite at isolated points of the interval of orthogonality*, Soviet Math. Doklady, 2 (1961), 1538-1541.

2 *The asymptotic behaviour of orthonormal polynomials at one-sided singular points of weighting functions (algebraic singularities)*, Soviet Math. Doklady, 6 (1965), 223-227.

Koschmieder, L.

1 *Turánshe und Forsythesche Funktionenfolgen*, Monatsh. Math., 64 (1960), 263-271.

Krall, H. L.

1 *On higher derivatives of orthogonal polynomials*, Bull. Amer. Math. Soc., 42 (1936), 867-870.

2 *Certain differential equations for Tchebycheff polynomials*, Duke Math. J., 4 (1938), 705-718.

3 *On derivatives of orthogonal polynomials II*, Bull. Amer. Math. Soc., 47 (1941), 261-264.

Krall, H. L. and Frink, O.

1 *A new class of orthogonal polynomials: the Bessel polynomials*, Trans. Amer. Math. Soc., 65 (1949), 100-115.

Krall, H. L. and Sheffer, I. M.

1 *Differential equations of infinite order for orthogonal polynomials*, Annali di Matematica pura ed applicata, (IV) LXXIV (1966), 135-172.

Krawtchouk, M.

1 *Sur une généralisation des polynomes d'Hermite*, C.R. Acad. Sci. Paris, 189 (1929), 620-622.

Krein, M. G.

  1 *The ideas of P. L. Čebyšev and A. A. Markov in the theory of limiting values of integrals and their further development*, Amer. Math. Soc. Transl., Series 2, 12 (1959), 1-122.

Krylov, V. I.

  1 *Approximate Calculation of Integrals*, Macmillan Co., N.Y., 1962.

Laguerre, E. N.

  1 *Sur l'integrale $\int_x^\infty x^{-1} e^{-x} dx$*, Bull. de la Société Math. de France, 7 (1879), 72-81.

Lancaster, O. E.

  1 *Orthogonal polynomials defined by difference equations*, Amer. J. Math., 63 (1941), 185-207.

Lauwerier, H. A.

  1 *On certain trigonometrical expansions*, J. Math. Mech., 8 (1959), 419-432.

Lee, P. A.

  1 *An integral representation and some summation formulas for the Hahn polynomials*, SIAM J. Appl. Math., 19 (1970), 266-272.

Legendre, A. M.

  1 *Recherches sur l'attraction des sphéroides homogènes*, Mémoires Math. Phys. présentés à l'Acad. Sci., 10 (1875), 411-434.

Leipnik, R.

  1 *Integral equations, biorthogonal expansions, and noise*, SIAM J., 7 (1959), 6-30.

Lesky, P.

  1 *Unendliche orthogonale Matrizen und Laguerresche Matrizen*, Monat. Math., 63 (1958), 59-83.
  2 *Die Übersetzung der klassischen orthogonalen Polynome in die Differenzenrechnung*, Monat. Math., 65 (1961), 1-26.
  3 *Erganzung zur Übersetzung der klassischen orthogonalen Polynome in die Differenzenrechnung*, Ibid., 66 (1962), 431-434.
  4 *Orthogonale Polynomsysteme als Lösungen Sturm-Liouvillscher Differenzengleichungen*, Ibid., 66 (1962), 203-214.
  5 *Über Polynomsysteme, die Sturm-Liouvillschen Differenzengleichungen genügen*, Math. Zeit., 78 (1962), 439-445.

Levit, R. J.

  1 *The zeros of the Hahn polynomials*, SIAM Rev., 9 (1967), 191-203.

Maki, D. P.

  1 *On constructing distribution functions: a bounded denumerable spectrum with n limit points*, Pacific J. Math., 22 (1967), 431-452.

2 *On constructing distribution functions*: *with applications to Lommel polynomials and Bessel functions*, Trans. Amer. Math. Soc., 130 (1968), 281-297.

3 *A note on recursively defined orthogonal polynomials*, Pacific J. Math., 28 (1969), 611-613.

4 *On the true interval of orthogonality*, Quart. J. Math., Oxford (2) 21 (1970), 61-65.

Markov, A. (Markoff)

1 *Démonstration de certaines inequalités de M. Tchébycheff*, Math. Annal. 24 (1884), 172-180.

2 *Differenzenrechnung*, Leipzig, 1896.

Meixner, J.

1 *Orthogonale Polynomsysteme mit einem besonderen Gestalt der erzeugenden Funktion*, J. London Math. Soc., 9 (1934), 6-13.

2 *Symmetric systems of orthogonal polynomials*, Arch. Rational Mech. Anal., 44 (1972), 69-75.

Morrison, N.

1 *Smoothing and extrapolation of time series by means of discrete Laguerre polynomials*, SIAM J. Appl. Math., 15 (1967), 516-538.

2 *Smoothing and extrapolation of continuous time series using Laguerre polynomials*, SIAM J. Appl. Math., 16 (1968), 1280-1304.

3 *Introduction to Sequential Smoothing and Prediction*, McGraw-Hill Book Co., N.Y., 1969.

Muckenhoupt, B.

1 *Poisson integrals for Hermite and Laguerre expansions*, Trans. Amer. Math. Soc., 139 (1969), 243-260.

2 *Conjugate functions for Laguerre expansions*, Trans. Amer. Math. Soc., 147 (1970), 403-418.

3 *Mean convergence of Hermite and Laguerre series, I, II*, Trans. Amer. Math. Soc., 147 (1970), 419-431; 433-460.

Muckenhoupt, B. and Stein, E. M.

1 *Classical expansions and their relation to conjugate harmonic functions*, Trans. Amer. Math. Soc., 118 (1965), 17-92.

Mukherjee, B. N. and Nanjundiah, T. S.

1 *On an inequality relating to Laguerre and Hermite polynomials*, Math. Student, 19 (1951), 47-48.

Natanson, I. P.

1 *Konstruktive Funktionentheorie*, Akademie-Verlag, Berlin, 1955.

Neuts, M. F. and Lambert, P. V.

1 *A limit theorem for orthogonal functions*, J. d'Analyse Math., X (1962/1963), 273-285.

Novikoff, A.

1 *On a special system of orthogonal polynomials*, Thesis, Stanford, 1954.

Palama, G.

1 *Polinomi più generali di altri classici e dei loro associati e relazioni tra essi funzioni di seconda specie*, Riv. Mat. Univ. Parma, 4 (1953), 363-383.

Pasternak, S.

1 *A generalization of the polynomial $F_n(x)$*, Philos. Mag. (7) 28 (1939), 209-226.

Peck, J. E. L.

1 *Polynomial curve fitting with constraint*, SIAM Review, 4 (1962), 135-141.

Pollaczek, F.

1 *Sur une generalization des polynomes de Legendre*, C.R. Acad. Sci. Paris, 228 (1949), 1363-1665.
2 *Systèmes de polynomes biorthogonaux à coefficients réels*, Ibid., 228 (1949), 1553-1556.
3 *Système de polynomes biorthogonaux qui généralisent les polynomes ultraspheriques*, Ibid., 228 (1949), 1998-2000.
4 *Familles de polynomes orthogonaux*, Ibid., 230 (1950), 36-37.
5 *Sur une famille de polynomes orthogonaux qui contient les polynomes d'Hermite et de Laguerre comme cas limites*, Ibid., 230 (1950), 1563-1565.
6 *Sur une famille de polynomes orthogonaux à quatre parametres*, Ibid., 230 (1950), 2254-2256.
7 *Sur une generalisation des polynomes de Jacobi*, Mémorial des Sciences Mathématiques, CXXI, Paris, 1956.

Pollard, H.

1 *The mean convergence of orthogonal series III*, Duke Math. J., 16 (1949), 189-191.

Prizva, G. I.

1 *Orthogonal polynomials associated with the Pascal probability distribution* [Ukranian], Visnik Kiivs'kogo Universitetu (Kiev), No. 8 (1966), 153-160.

Rainville, E. D.

1 *Special Functions*, Macmillan Co., N.Y., 1960.

Rees, C. J.

1 *Elliptic orthogonal polynomials*, Duke Math. J., 12 (1945), 173-187.

Riesz, M.

1 *Sur le problème des moments. Troisième Note*, Arkiv för matematik, astronomi och fysik, 17 (16) (1923), 52 pp.
2 *Sur le problème des moments et le théorème de Parseval correspondant*, Acta Litterarum ac Scientiarum (Szeged), 1 (1922-1923), 209-225.

Rodrigues, O.

1 *Mémoire sur l'attraction des spheroides*, Corresp. sur l'Ecole Royale Polytech. III (1816), 361-385.

Romanovsky, V.

1 *Generalization of some types of the frequency curves of Professor Pearson*, Biometrika, 16 (1924), 106-117.
2 *Sur quelques classes nouvelles de polynomes orthogonaux*, C.R. Acad. Sci., Paris, 188 (1929), 1023-1025.

Rossum, H. van

1 *Contiguous orthogonal systems*, Koninkl. Nederl. Akad. von Wetenschappen - Amsterdam, Proc., A, 63 (1960), 323-332.
2 *Lacunary orthogonal polynomials, Ibid.*, 68 (1965), 55-63.
3 *Totally positive polynomials, Ibid.*, 68 (1965), 305-315.
4 *A note on the location of the zeros of generalized Bessel polynomials and totally positive polynomials*, Niew Archief. voor Wiskunde (2), XVII (1969), 1-8.

Rudin, W.

1 *Principles of Mathematical Analysis*, McGraw-Hill Book Co., 2nd Ed., 1964.

Sallay, M.

1 *Über "lückenhafte" Orthogonalpolynomsysteme*, Studia Sci. Math. Hungar., 4 (1969), 371-377.

Salzer, H. E.

1 *Orthogonal polynomials arising in the numerical evaluation of inverse Laplace transforms*, Math. Tables Aids Comp., 9 (1955), 164-177.

Sansone, G.

1 *Orthogonal Functions*, Interscience Publ., N.Y., 1959.

Seall, R. and Wetzel, M.

1 *Some connections between continued fractions and convex sets*, Pacific J. Math., 9 (1959), 861-873.
2 *Quadratic forms and chain sequences*, Proc. Amer. Math. Soc., 15 (1964), 729-734.

Shao, T. S., Chen, T. C. and Frank, R. M.

1 *Tables of zeros and Gaussian weights of certain associated Laguerre polynomials and the related generalized Hermite polynomials*, Math. Comp., XVIII (1964), 598-616.

Sheffer, I. M.

1 *Concerning Appell sets and associated linear functional equations*, Duke Math. J., 3 (1937), 593-609.
2 *Some properties of polynomial sets of type zero*, Duke Math. J., 5 (1939), 590-622.

Sherman, J.

1 *On the numerators of the convergents of the Stieltjes continued fractions*, Trans. Amer. Math. Soc., 35 (1933), 64-87.

Shohat, J. A.

1 *Théorie générale des polynomes orthogonaux de Tchebychef*, Mémoires des Sciences Mathématiques, 66 (1934), 1-68.
2 *The relation of the classical orthogonal polynomials to the polynomials of Appell*, Amer. J. Math., 58 (1936), 453-464.
3 *Sur les polynomes orthogonaux généralisees*, C.R. Acad. Sci., Paris, 207 (1938), 556-558.

Shohat, J. A., Hille, E. and Walsh, J. L.

1 *A Bibliography on Orthogonal Polynomials*, Bull. National Research Council (U.S.A.), 103 (1940).

Shohat, J. A. and Tamarkin, J. D.

1 *The Problem of Moments*, Mathematical Surveys No. 1, Amer. Math. Soc., N.Y., 1943, 1950.

Shore, S. D.

1 *On the second order differential equation which has orthogonal polynomial solutions*, Bull. Calcutta Math. Soc., 56 (1964), 195-198.

Skovgaard, H.

1 *On inequalities of the Turán type*, Math. Scand., 2 (1954), 65-73.

Snyder, M. A.

1 *Chebyshev Methods in Numerical Approximation*, Prentice-Hall, Englewood Cliffs, 1966.

Sonine, N. J.

1 *Recherches sur les fonctions cylindriques et le développment des fonctions continues en series*, Math. Ann., 16 (1880), 1-80.

Stiefel, E. L.

1 *Kernel polynomials in linear algebra and their numerical applications*, Nat. Bur. Standards Appl. Math. Series, 49 (1958), 1-22.

Stieltjes, T. J.

1 *Sur quelques intégrals definies et leur développement en fractions continues*, Quart. J. Math., 24 (1890), 370-382; Oeuvres, Vol. 2, Noordhoff, Groningen, 1918, 378-394.
2 *Recherches sur les fractions continues*, Annales de la Faculté des Sciences de Toulouse, 8 (1894), J1-122; 9 (1895), A1-47; Oeuvres, Vol. 2, 398-566.

Stone, M. H.

1 *Linear Transformations in Hilbert Space*, Amer. Math. Soc. Colloq. Publ., Vol. 15, Amer. Math. Soc., N.Y., 1932.

Stroud, A. H. and Secrest, D.

1 *Approximate integration formulas for certain spherically symmetric regions*, Math. Comp., 17 (1963), 105-135.
2 *Gaussian quadrature formulas*, Prentice-Hall, Engelwood Cliffs, 1966.

Struble, G. W.

1 *Orthogonal polynomials: Variable Signed Weight Functions*, Thesis, Wisconsin, 1961.

Szász, O.

1 *Identities and inequalities concerning orthogonal polynomials and Bessel functions*, J. d'Analyse Math. I (1951), 116-134.

Szegö, G.

1 *Ein Beitrag zur theorie der Polynomie von Laguerre und Jacobi*, Math. Zeit., 1 (1918), 341-356.
2 *Über die Entwicklung einer analytische Funktion nach den Polynomen eines Orthogonalsystems*, Math. Ann., 82 (1921), 188-212.
3 *Ein Beitrag zur Theorie der Thetafunktionen*, Sitzungs. Pruss. Akad. Wissen., Phys.-math. Klasse, 1926, 242-252.
4 *Über gewisse orthogonale Polynome die zu einer oszillierenden Belegungsfunktion gehören*, Math. Ann., 110 (1935), 501-513.
5 *Orthogonal Polynomials*, Amer. Math. Soc. Colloq. Publ. Vol. 23, Amer. Math. Soc., N.Y., 1939, 3rd Edition 1967.
6 *On an inequality of P. Turán concerning the Legendre polynomials*, Bull. Amer. Math. Soc., 54 (1948), 401-405.
7 *On certain special sets of orthogonal polynomials*, Proc. Amer. Math. Soc., 1 (1950), 731-737.
8 *Recent advances and open questions on the asymptotic expansions of orthogonal polynomials*, SIAM J., 7 (1959), 311-315.
9 *An equality for Jacobi polynomials*, in Studies in Mathematical Analysis and Related Topics, Ed. by G. Szegö et al., Stanford U. Press, Stanford, 1962, 392-398.

Tchebichef, P. L. (Čebyšev, Tchebycheff, etc.)

1 *Sur le développment des fonctions à une seule variable*. Bull. phys.-math. Acad. Imp. Sc. St. Petersburg, I (1859), 193-200; Oeuvres, Vol. I, 539-560.
2 *Sur l'interpolation*, Zapiski Akademii Nauk, 4, Supplement 5, 1864; Oeuvres, Vol. 1, 539-560.
3 *Sur les valeurs limites des intégrales*, J. Math. Pures Appl., 19 (1874), 157-160; Oeuvres, Vol. 2, 181-185.

Toscano, L.

1 *Polinomi associati a polinomi classici*, Rivista Mat. Univ. Parma, 4 (1953), 387-402.

Touchard, J.

1 *Nombres exponentiels et nombres de Bernoulli*, Canad. J. Math., 8 (1956), 305-320.

Trench, W. F.

1 *Proof of a conjecture of Askey on orthogonal expansions with positive coefficients*, Bull. Amer. Math. Soc., 81 (1975), 954-956.

Tricomi, F.

1 *Equazioni differentiali*, Torino, 1948.
2 *A class of non-orthogonal polynomials related to those of Laguerre*, J. d'Analyse Math., I (1951), 209-231.
3 *Vorlesungen über orthogonalreihen*, Springer-Verlag, Berlin, 1955.

Ullman, J. L.

1 *A class of weight functions that admit Tchebycheff quadrature*, Michigan Math. J., 12 (1966), 417-423.
2 *On the regular behavior of orthogonal polynomials*, Proc. London Math. Soc., (3rd Series) XXIV (1972), 119-148.

Wall, H. S.

1 *A continued fraction related to some partition formulas of Euler*, Amer. Math. Monthly, 48 (1941), 102-108.
2 *Analytic Theory of Continued Fractions*, D. van Nostrand Co., N.Y., 1948.

Weber, M. and Erdélyi, A.

1 *On the finite difference analogue of Rodrigues' formula*, Amer. Math. Monthly, 59 (1952), 163-168.

Weisner, L.

1 *Problem 4766*, Amer. Math. Monthly, 64 (1957), 747.

Wendroff, B.

1 *On orthogonal polynomials*, Proc. Amer. Math. Soc., 12 (1961), 554-555.

Weyl, H.

1 *The Theory of Groups and Quantum Mechanics*, London, 1931, 2nd Ed., reprinted, Dover Publ., N.Y.

Whittaker, E. T. and Watson, G. N.

1 *A course of Modern Analysis*, Cambridge U. Press, Cambridge, 4th Ed., 1927.

Wigert, S.

1 *Sur les polynomes orthogonaux et l'approximation des fonctions continues*, Arkiv för matematik, astronomi och fysik, 17 (1923), No. 18, 15 pp.

Wilson, M. W.

1 *On the Hahn polynomials*, SIAM J. Math. Anal., 1 (1970), 131-139.

2 *Convergence properties of discrete analogs of orthogonal polynomials*, Computing, 5 (1970), 1-5.

3 *Nonnegative expansions of polynomials*, Proc. Amer. Math. Soc., 24 (1970), 100-102.

4 *On a new discrete analog of the Legendre polynomials*, SIAM J. Math. 3 (1972), 157-169.

Wintner, A.

1 *The moment problem of enumerating distributions*, Duke Math. J., 12 (1945), 23-25.

Wyman, M. and Moser, L.

1 *On some polynomials of Touchard*, Canad. J. Math., 8 (1956), 321-322.

Wynn, P.

1 *A general system of orthogonal polynomials*, Quart. J. Math. (Oxford), 2 (1967), 81-96.

Young, T. Y.

1 *Binomial weighted orthogonal polynomials*, J. Assoc. Comput. Mach., 14 (1967), 120-127.

Inönü, E.

1 *Orthogonality of a set of polynomials encountered in neutron-transport and radiature-transfer theories*, J. Math. Physics, 11 (1970), 568-577.

Index

243

A CATALOG OF SELECTED
DOVER BOOKS
IN SCIENCE AND MATHEMATICS

Astronomy

BURNHAM'S CELESTIAL HANDBOOK, Robert Burnham, Jr. Thorough guide to the stars beyond our solar system. Exhaustive treatment. Alphabetical by constellation: Andromeda to Cetus in Vol. 1; Chamaeleon to Orion in Vol. 2; and Pavo to Vulpecula in Vol. 3. Hundreds of illustrations. Index in Vol. 3. 2,000pp. 6⅛ x 9¼.

Vol. I: 0-486-23567-X
Vol. II: 0-486-23568-8
Vol. III: 0-486-23673-0

EXPLORING THE MOON THROUGH BINOCULARS AND SMALL TELE-SCOPES, Ernest H. Cherrington, Jr. Informative, profusely illustrated guide to locating and identifying craters, rills, seas, mountains, other lunar features. Newly revised and updated with special section of new photos. Over 100 photos and diagrams. 240pp. 8¼ x 11. 0-486-24491-1

THE EXTRATERRESTRIAL LIFE DEBATE, 1750–1900, Michael J. Crowe. First detailed, scholarly study in English of the many ideas that developed from 1750 to 1900 regarding the existence of intelligent extraterrestrial life. Examines ideas of Kant, Herschel, Voltaire, Percival Lowell, many other scientists and thinkers. 16 illustrations. 704pp. 5⅜ x 8½. 0-486-40675-X

THEORIES OF THE WORLD FROM ANTIQUITY TO THE COPERNICAN REVOLUTION, Michael J. Crowe. Newly revised edition of an accessible, enlightening book re-creates the change from an earth-centered to a sun-centered conception of the solar system. 242pp. 5⅜ x 8½. 0-486-41444-2

ARISTARCHUS OF SAMOS: The Ancient Copernicus, Sir Thomas Heath. Heath's history of astronomy ranges from Homer and Hesiod to Aristarchus and includes quotes from numerous thinkers, compilers, and scholasticists from Thales and Anaximander through Pythagoras, Plato, Aristotle, and Heraclides. 34 figures. 448pp. 5⅜ x 8½. 0-486-43886-4

A COMPLETE MANUAL OF AMATEUR ASTRONOMY: TOOLS AND TECHNIQUES FOR ASTRONOMICAL OBSERVATIONS, P. Clay Sherrod with Thomas L. Koed. Concise, highly readable book discusses: selecting, setting up and maintaining a telescope; amateur studies of the sun; lunar topography and occultations; observations of Mars, Jupiter, Saturn, the minor planets and the stars; an introduction to photoelectric photometry; more. 1981 ed. 124 figures. 25 halftones. 37 tables. 335pp. 6½ x 9¼. 0-486-42820-8

AMATEUR ASTRONOMER'S HANDBOOK, J. B. Sidgwick. Timeless, comprehensive coverage of telescopes, mirrors, lenses, mountings, telescope drives, micrometers, spectroscopes, more. 189 illustrations. 576pp. 5⅜ x 8¼. (Available in U.S. only.) 0-486-24034-7

STAR LORE: Myths, Legends, and Facts, William Tyler Olcott. Captivating retellings of the origins and histories of ancient star groups include Pegasus, Ursa Major, Pleiades, signs of the zodiac, and other constellations. "Classic."—Sky & Telescope. 58 illustrations. 544pp. 5⅜ x 8½. 0-486-43581-4

Chemistry

THE SCEPTICAL CHYMIST: THE CLASSIC 1661 TEXT, Robert Boyle. Boyle defines the term "element," asserting that all natural phenomena can be explained by the motion and organization of primary particles. 1911 ed. viii+232pp. $5^3/_8$ x $8^1/_2$.
0-486-42825-7

RADIOACTIVE SUBSTANCES, Marie Curie. Here is the celebrated scientist's doctoral thesis, the prelude to her receipt of the 1903 Nobel Prize. Curie discusses establishing atomic character of radioactivity found in compounds of uranium and thorium; extraction from pitchblende of polonium and radium; isolation of pure radium chloride; determination of atomic weight of radium; plus electric, photographic, luminous, heat, color effects of radioactivity. ii+94pp. $5^3/_8$ x $8^1/_2$.
0-486-42550-9

CHEMICAL MAGIC, Leonard A. Ford. Second Edition, Revised by E. Winston Grundmeier. Over 100 unusual stunts demonstrating cold fire, dust explosions, much more. Text explains scientific principles and stresses safety precautions. 128pp. $5^3/_8$ x $8^1/_2$.
0-486-67628-5

MOLECULAR THEORY OF CAPILLARITY, J. S. Rowlinson and B. Widom. History of surface phenomena offers critical and detailed examination and assessment of modern theories, focusing on statistical mechanics and application of results in mean-field approximation to model systems. 1989 edition. 352pp. $5^3/_8$ x $8^1/_2$.
0-486-42544-4

CHEMICAL AND CATALYTIC REACTION ENGINEERING, James J. Carberry. Designed to offer background for managing chemical reactions, this text examines behavior of chemical reactions and reactors; fluid-fluid and fluid-solid reaction systems; heterogeneous catalysis and catalytic kinetics; more. 1976 edition. 672pp. $6^1/_8$ x $9^1/_4$.
0-486-41736-0 $31.95

ELEMENTS OF CHEMISTRY, Antoine Lavoisier. Monumental classic by founder of modern chemistry in remarkable reprint of rare 1790 Kerr translation. A must for every student of chemistry or the history of science. 539pp. $5^3/_8$ x $8^1/_2$.
0-486-64624-6

MOLECULES AND RADIATION: An Introduction to Modern Molecular Spectroscopy. Second Edition, Jeffrey I. Steinfeld. This unified treatment introduces upper-level undergraduates and graduate students to the concepts and the methods of molecular spectroscopy and applications to quantum electronics, lasers, and related optical phenomena. 1985 edition. 512pp. $5^3/_8$ x $8^1/_2$.
0-486-44152-0

A SHORT HISTORY OF CHEMISTRY, J. R. Partington. Classic exposition explores origins of chemistry, alchemy, early medical chemistry, nature of atmosphere, theory of valency, laws and structure of atomic theory, much more. 428pp. $5^3/_8$ x $8^1/_2$. (Available in U.S. only.)
0-486-65977-1

GENERAL CHEMISTRY, Linus Pauling. Revised 3rd edition of classic first-year text by Nobel laureate. Atomic and molecular structure, quantum mechanics, statistical mechanics, thermodynamics correlated with descriptive chemistry. Problems. 992pp. $5^3/_8$ x $8^1/_2$.
0-486-65622-5

ELECTRON CORRELATION IN MOLECULES, S. Wilson. This text addresses one of theoretical chemistry's central problems. Topics include molecular electronic structure, independent electron models, electron correlation, the linked diagram theorem, and related topics. 1984 edition. 304pp. $5^3/_8$ x $8^1/_2$.
0-486-45879-2

Engineering

DE RE METALLICA, Georgius Agricola. The famous Hoover translation of greatest treatise on technological chemistry, engineering, geology, mining of early modern times (1556). All 289 original woodcuts. 638pp. $6\frac{3}{4}$ x 11.   0-486-60006-8

FUNDAMENTALS OF ASTRODYNAMICS, Roger Bate et al. Modern approach developed by U.S. Air Force Academy. Designed as a first course. Problems, exercises. Numerous illustrations. 455pp. $5\frac{5}{8}$ x $8\frac{1}{2}$.   0-486-60061-0

DYNAMICS OF FLUIDS IN POROUS MEDIA, Jacob Bear. For advanced students of ground water hydrology, soil mechanics and physics, drainage and irrigation engineering and more. 335 illustrations. Exercises, with answers. 784pp. $6\frac{1}{8}$ x $9\frac{1}{4}$.   0-486-65675-6

THEORY OF VISCOELASTICITY (SECOND EDITION), Richard M. Christensen. Complete consistent description of the linear theory of the viscoelastic behavior of materials. Problem-solving techniques discussed. 1982 edition. 29 figures. xiv+364pp. $6\frac{1}{8}$ x $9\frac{1}{4}$.   0-486-42880-X

MECHANICS, J. P. Den Hartog. A classic introductory text or refresher. Hundreds of applications and design problems illuminate fundamentals of trusses, loaded beams and cables, etc. 334 answered problems. 462pp. $5\frac{5}{8}$ x $8\frac{1}{2}$.   0-486-60754-2

MECHANICAL VIBRATIONS, J. P. Den Hartog. Classic textbook offers lucid explanations and illustrative models, applying theories of vibrations to a variety of practical industrial engineering problems. Numerous figures. 233 problems, solutions. Appendix. Index. Preface. 436pp. $5\frac{5}{8}$ x $8\frac{1}{2}$.   0-486-64785-4

STRENGTH OF MATERIALS, J. P. Den Hartog. Full, clear treatment of basic material (tension, torsion, bending, etc.) plus advanced material on engineering methods, applications. 350 answered problems. 323pp. $5\frac{5}{8}$ x $8\frac{1}{2}$.   0-486-60755-0

A HISTORY OF MECHANICS, René Dugas. Monumental study of mechanical principles from antiquity to quantum mechanics. Contributions of ancient Greeks, Galileo, Leonardo, Kepler, Lagrange, many others. 671pp. $5\frac{5}{8}$ x $8\frac{1}{2}$.   0-486-65632-2

STABILITY THEORY AND ITS APPLICATIONS TO STRUCTURAL MECHANICS, Clive L. Dym. Self-contained text focuses on Koiter postbuckling analyses, with mathematical notions of stability of motion. Basing minimum energy principles for static stability upon dynamic concepts of stability of motion, it develops asymptotic buckling and postbuckling analyses from potential energy considerations, with applications to columns, plates, and arches. 1974 ed. 208pp. $5\frac{5}{8}$ x $8\frac{1}{2}$.   0-486-42541-X

BASIC ELECTRICITY, U.S. Bureau of Naval Personnel. Originally a training course; best nontechnical coverage. Topics include batteries, circuits, conductors, AC and DC, inductance and capacitance, generators, motors, transformers, amplifiers, etc. Many questions with answers. 349 illustrations. 1969 edition. 448pp. $6\frac{1}{2}$ x $9\frac{1}{4}$.   0-486-20973-3

ROCKETS, Robert Goddard. Two of the most significant publications in the history of rocketry and jet propulsion: "A Method of Reaching Extreme Altitudes" (1919) and "Liquid Propellant Rocket Development" (1936). 128pp. 5³⁄₈ x 8¹⁄₂.   0-486-42537-1

STATISTICAL MECHANICS: PRINCIPLES AND APPLICATIONS, Terrell L. Hill. Standard text covers fundamentals of statistical mechanics, applications to fluctuation theory, imperfect gases, distribution functions, more. 448pp. 5³⁄₈ x 8¹⁄₂.   0-486-65390-0

ENGINEERING AND TECHNOLOGY 1650–1750: ILLUSTRATIONS AND TEXTS FROM ORIGINAL SOURCES, Martin Jensen. Highly readable text with more than 200 contemporary drawings and detailed engravings of engineering projects dealing with surveying, leveling, materials, hand tools, lifting equipment, transport and erection, piling, bailing, water supply, hydraulic engineering, and more. Among the specific projects outlined-transporting a 50-ton stone to the Louvre, erecting an obelisk, building timber locks, and dredging canals. 207pp. 8³⁄₈ x 11¹⁄₄.   0-486-42232-1

THE VARIATIONAL PRINCIPLES OF MECHANICS, Cornelius Lanczos. Graduate level coverage of calculus of variations, equations of motion, relativistic mechanics, more. First inexpensive paperbound edition of classic treatise. Index. Bibliography. 418pp. 5³⁄₈ x 8¹⁄₂.   0-486-65067-7

PROTECTION OF ELECTRONIC CIRCUITS FROM OVERVOLTAGES, Ronald B. Standler. Five-part treatment presents practical rules and strategies for circuits designed to protect electronic systems from damage by transient overvoltages. 1989 ed. xxiv+434pp. 6¹⁄₈ x 9¹⁄₄.   0-486-42552-5

ROTARY WING AERODYNAMICS, W. Z. Stepniewski. Clear, concise text covers aerodynamic phenomena of the rotor and offers guidelines for helicopter performance evaluation. Originally prepared for NASA. 537 figures. 640pp. 6¹⁄₈ x 9¹⁄₄. 0-486-64647-5

INTRODUCTION TO SPACE DYNAMICS, William Tyrrell Thomson. Comprehensive, classic introduction to space-flight engineering for advanced undergraduate and graduate students. Includes vector algebra, kinematics, transformation of coordinates. Bibliography. Index. 352pp. 5³⁄₈ x 8¹⁄₂.   0-486-65113-4

HISTORY OF STRENGTH OF MATERIALS, Stephen P. Timoshenko. Excellent historical survey of the strength of materials with many references to the theories of elasticity and structure. 245 figures. 452pp. 5³⁄₈ x 8¹⁄₂.   0-486-61187-6

ANALYTICAL FRACTURE MECHANICS, David J. Unger. Self-contained text supplements standard fracture mechanics texts by focusing on analytical methods for determining crack-tip stress and strain fields. 336pp. 6¹⁄₈ x 9¹⁄₄.   0-486-41737-9

STATISTICAL MECHANICS OF ELASTICITY, J. H. Weiner. Advanced, self-contained treatment illustrates general principles and elastic behavior of solids. Part 1, based on classical mechanics, studies thermoelastic behavior of crystalline and polymeric solids. Part 2, based on quantum mechanics, focuses on interatomic force laws, behavior of solids, and thermally activated processes. For students of physics and chemistry and for polymer physicists. 1983 ed. 96 figures. 496pp. 5³⁄₈ x 8¹⁄₂.   0-486-42260-7

Mathematics

FUNCTIONAL ANALYSIS (Second Corrected Edition), George Bachman and Lawrence Narici. Excellent treatment of subject geared toward students with background in linear algebra, advanced calculus, physics and engineering. Text covers introduction to inner-product spaces, normed, metric spaces, and topological spaces; complete orthonormal sets, the Hahn-Banach Theorem and its consequences, and many other related subjects. 1966 ed. 544pp. $6^1/_8$ x $9^1/_4$. 0-486-40251-7

DIFFERENTIAL MANIFOLDS, Antoni A. Kosinski. Introductory text for advanced undergraduates and graduate students presents systematic study of the topological structure of smooth manifolds, starting with elements of theory and concluding with method of surgery. 1993 edition. 288pp. $5^3/_8$ x $8^1/_2$. 0-486-46244-7

VECTOR AND TENSOR ANALYSIS WITH APPLICATIONS, A. I. Borisenko and I. E. Tarapov. Concise introduction. Worked-out problems, solutions, exercises. 257pp. $5^5/_8$ x $8^1/_4$. 0-486-63833-2

AN INTRODUCTION TO ORDINARY DIFFERENTIAL EQUATIONS, Earl A. Coddington. A thorough and systematic first course in elementary differential equations for undergraduates in mathematics and science, with many exercises and problems (with answers). Index. 304pp. $5^3/_8$ x $8^1/_2$. 0-486-65942-9

FOURIER SERIES AND ORTHOGONAL FUNCTIONS, Harry F. Davis. An incisive text combining theory and practical example to introduce Fourier series, orthogonal functions and applications of the Fourier method to boundary-value problems. 570 exercises. Answers and notes. 416pp. $5^3/_8$ x $8^1/_2$. 0-486-65973-9

COMPUTABILITY AND UNSOLVABILITY, Martin Davis. Classic graduate-level introduction to theory of computability, usually referred to as theory of recurrent functions. New preface and appendix. 288pp. $5^3/_8$ x $8^1/_2$. 0-486-61471-9

AN INTRODUCTION TO MATHEMATICAL ANALYSIS, Robert A. Rankin. Dealing chiefly with functions of a single real variable, this text by a distinguished educator introduces limits, continuity, differentiability, integration, convergence of infinite series, double series, and infinite products. 1963 edition. 624pp. $5^3/_8$ x $8^1/_2$. 0-486-46251-X

METHODS OF NUMERICAL INTEGRATION (SECOND EDITION), Philip J. Davis and Philip Rabinowitz. Requiring only a background in calculus, this text covers approximate integration over finite and infinite intervals, error analysis, approximate integration in two or more dimensions, and automatic integration. 1984 edition. 624pp. $5^3/_8$ x $8^1/_2$. 0-486-45339-1

INTRODUCTION TO LINEAR ALGEBRA AND DIFFERENTIAL EQUATIONS, John W. Dettman. Excellent text covers complex numbers, determinants, orthonormal bases, Laplace transforms, much more. Exercises with solutions. Undergraduate level. 416pp. $5^3/_8$ x $8^1/_2$. 0-486-65191-6

RIEMANN'S ZETA FUNCTION, H. M. Edwards. Superb, high-level study of landmark 1859 publication entitled "On the Number of Primes Less Than a Given Magnitude" traces developments in mathematical theory that it inspired. xiv+315pp. $5^3/_8$ x $8^1/_2$. 0-486-41740-9

CALCULUS OF VARIATIONS WITH APPLICATIONS, George M. Ewing. Applications-oriented introduction to variational theory develops insight and promotes understanding of specialized books, research papers. Suitable for advanced undergraduate/graduate students as primary, supplementary text. 352pp. 5⅜ x 8½.
0-486-64856-7

MATHEMATICIAN'S DELIGHT, W. W. Sawyer. "Recommended with confidence" by *The Times Literary Supplement,* this lively survey was written by a renowned teacher. It starts with arithmetic and algebra, gradually proceeding to trigonometry and calculus. 1943 edition. 240pp. 5⅜ x 8½.
0-486-46240-4

ADVANCED EUCLIDEAN GEOMETRY, Roger A. Johnson. This classic text explores the geometry of the triangle and the circle, concentrating on extensions of Euclidean theory, and examining in detail many relatively recent theorems. 1929 edition. 336pp. 5⅜ x 8½.
0-486-46237-4

COUNTEREXAMPLES IN ANALYSIS, Bernard R. Gelbaum and John M. H. Olmsted. These counterexamples deal mostly with the part of analysis known as "real variables." The first half covers the real number system, and the second half encompasses higher dimensions. 1962 edition. xxiv+198pp. 5⅜ x 8½.
0-486-42875-3

CATASTROPHE THEORY FOR SCIENTISTS AND ENGINEERS, Robert Gilmore. Advanced-level treatment describes mathematics of theory grounded in the work of Poincaré, R. Thom, other mathematicians. Also important applications to problems in mathematics, physics, chemistry and engineering. 1981 edition. References. 28 tables. 397 black-and-white illustrations. xvii + 666pp. 6⅛ x 9¼.
0-486-67539-4

COMPLEX VARIABLES: Second Edition, Robert B. Ash and W. P. Novinger. Suitable for advanced undergraduates and graduate students, this newly revised treatment covers Cauchy theorem and its applications, analytic functions, and the prime number theorem. Numerous problems and solutions. 2004 edition. 224pp. 6½ x 9¼.
0-486-46250-1

NUMERICAL METHODS FOR SCIENTISTS AND ENGINEERS, Richard Hamming. Classic text stresses frequency approach in coverage of algorithms, polynomial approximation, Fourier approximation, exponential approximation, other topics. Revised and enlarged 2nd edition. 721pp. 5⅜ x 8½.
0-486-65241-6

INTRODUCTION TO NUMERICAL ANALYSIS (2nd Edition), F. B. Hildebrand. Classic, fundamental treatment covers computation, approximation, interpolation, numerical differentiation and integration, other topics. 150 new problems. 669pp. 5⅜ x 8½.
0-486-65363-3

MARKOV PROCESSES AND POTENTIAL THEORY, Robert M. Blumental and Ronald K. Getoor. This graduate-level text explores the relationship between Markov processes and potential theory in terms of excessive functions, multiplicative functionals and subprocesses, additive functionals and their potentials, and dual processes. 1968 edition. 320pp. 5⅜ x 8½.
0-486-46263-3

ABSTRACT SETS AND FINITE ORDINALS: An Introduction to the Study of Set Theory, G. B. Keene. This text unites logical and philosophical aspects of set theory in a manner intelligible to mathematicians without training in formal logic and to logicians without a mathematical background. 1961 edition. 112pp. 5⅜ x 8½.
0-486-46249-8

INTRODUCTORY REAL ANALYSIS, A.N. Kolmogorov, S. V. Fomin. Translated by Richard A. Silverman. Self-contained, evenly paced introduction to real and functional analysis. Some 350 problems. 403pp. 5⅜ x 8½. 0-486-61226-0

APPLIED ANALYSIS, Cornelius Lanczos. Classic work on analysis and design of finite processes for approximating solution of analytical problems. Algebraic equations, matrices, harmonic analysis, quadrature methods, much more. 559pp. 5⅜ x 8½. 0-486-65656-X

AN INTRODUCTION TO ALGEBRAIC STRUCTURES, Joseph Landin. Superb self-contained text covers "abstract algebra": sets and numbers, theory of groups, theory of rings, much more. Numerous well-chosen examples, exercises. 247pp. 5⅜ x 8½.
0-486-65940-2

QUALITATIVE THEORY OF DIFFERENTIAL EQUATIONS, V. V. Nemytskii and V.V. Stepanov. Classic graduate-level text by two prominent Soviet mathematicians covers classical differential equations as well as topological dynamics and ergodic theory. Bibliographies. 523pp. 5⅜ x 8½. 0-486-65954-2

THEORY OF MATRICES, Sam Perlis. Outstanding text covering rank, nonsingularity and inverses in connection with the development of canonical matrices under the relation of equivalence, and without the intervention of determinants. Includes exercises. 237pp. 5⅜ x 8½. 0-486-66810-X

INTRODUCTION TO ANALYSIS, Maxwell Rosenlicht. Unusually clear, accessible coverage of set theory, real number system, metric spaces, continuous functions, Riemann integration, multiple integrals, more. Wide range of problems. Undergraduate level. Bibliography. 254pp. 5⅜ x 8½. 0-486-65038-3

MODERN NONLINEAR EQUATIONS, Thomas L. Saaty. Emphasizes practical solution of problems; covers seven types of equations. ". . . a welcome contribution to the existing literature. . . ."—*Math Reviews*. 490pp. 5⅜ x 8½. 0-486-64232-1

MATRICES AND LINEAR ALGEBRA, Hans Schneider and George Phillip Barker. Basic textbook covers theory of matrices and its applications to systems of linear equations and related topics such as determinants, eigenvalues and differential equations. Numerous exercises. 432pp. 5⅜ x 8½. 0-486-66014-1

LINEAR ALGEBRA, Georgi E. Shilov. Determinants, linear spaces, matrix algebras, similar topics. For advanced undergraduates, graduates. Silverman translation. 387pp. 5⅜ x 8½. 0-486-63518-X

MATHEMATICAL METHODS OF GAME AND ECONOMIC THEORY: Revised Edition, Jean-Pierre Aubin. This text begins with optimization theory and convex analysis, followed by topics in game theory and mathematical economics, and concluding with an introduction to nonlinear analysis and control theory. 1982 edition. 656pp. 6⅛ x 9¼.
0-486-46265-X

SET THEORY AND LOGIC, Robert R. Stoll. Lucid introduction to unified theory of mathematical concepts. Set theory and logic seen as tools for conceptual understanding of real number system. 496pp. 5⅜ x 8¼. 0-486-63829-4

CATALOG OF DOVER BOOKS

A TREATISE ON ELECTRICITY AND MAGNETISM, James Clerk Maxwell. Important foundation work of modern physics. Brings to final form Maxwell's theory of electromagnetism and rigorously derives his general equations of field theory. 1,084pp. 5⅜ x 8½. Two-vol. set.   Vol. I: 0-486-60636-8   Vol. II: 0-486-60637-6

MATHEMATICS FOR PHYSICISTS, Philippe Dennery and Andre Krzywicki. Superb text provides math needed to understand today's more advanced topics in physics and engineering. Theory of functions of a complex variable, linear vector spaces, much more. Problems. 1967 edition. 400pp. 6½ x 9¼.   0-486-69193-4

INTRODUCTION TO QUANTUM MECHANICS WITH APPLICATIONS TO CHEMISTRY, Linus Pauling & E. Bright Wilson, Jr. Classic undergraduate text by Nobel Prize winner applies quantum mechanics to chemical and physical problems. Numerous tables and figures enhance the text. Chapter bibliographies. Appendices. Index. 468pp. 5⅜ x 8½.   0-486-64871-0

METHODS OF THERMODYNAMICS, Howard Reiss. Outstanding text focuses on physical technique of thermodynamics, typical problem areas of understanding, and significance and use of thermodynamic potential. 1965 edition. 238pp. 5⅜ x 8½.
0-486-69445-3

THE ELECTROMAGNETIC FIELD, Albert Shadowitz. Comprehensive under- graduate text covers basics of electric and magnetic fields, builds up to electromagnetic theory. Also related topics, including relativity. Over 900 problems. 768pp. 5⅜ x 8¼.
0-486-65660-8

GREAT EXPERIMENTS IN PHYSICS: FIRSTHAND ACCOUNTS FROM GALILEO TO EINSTEIN, Morris H. Shamos (ed.). 25 crucial discoveries: Newton's laws of motion, Chadwick's study of the neutron, Hertz on electromagnetic waves, more. Original accounts clearly annotated. 370pp. 5⅜ x 8½.   0-486-25346-5

EINSTEIN'S LEGACY, Julian Schwinger. A Nobel Laureate relates fascinating story of Einstein and development of relativity theory in well-illustrated, nontechnical volume. Subjects include meaning of time, paradoxes of space travel, gravity and its effect on light, non-Euclidean geometry and curving of space-time, impact of radio astronomy and space-age discoveries, and more. 189 b/w illustrations. xiv+250pp. 8⅜ x 9¼.   0-486-41974-6

THE VARIATIONAL PRINCIPLES OF MECHANICS, Cornelius Lanczos. Philosophic, less formalistic approach to analytical mechanics offers model of clear, scholarly exposition at graduate level with coverage of basics, calculus of variations, principle of virtual work, equations of motion, more. 418pp. 5⅜ x 8½.   0-486-65067-7

Paperbound unless otherwise indicated. Available at your book dealer, online at www.doverpublications.com, or by writing to Dept. GI, Dover Publications, Inc., 31 East 2nd Street, Mineola, NY 11501. For current price information or for free catalogues (please indicate field of interest), write to Dover Publications or log on to www.doverpublications.com and see every Dover book in print. Dover publishes more than 400 books each year on science, elementary and advanced mathematics, biology, music, art, literary history, social sciences, and other areas.